Natural
Environment
Research
Council

E8

Angling and wildlife in fresh waters

ITE symposium no. 19

Proceedings of a symposium organized by
the Scottish Freshwater Group and
the British Ecological Society
University of Stirling
30 October 1985

Edited by

P S MAITLAND
Institute of Terrestrial Ecology
Bush Estate
PENICUIK
Midlothian

and

A K TURNER
Department of Zoology
University of Glasgow
GLASGOW

INSTITUTE OF TERRESTRIAL ECOLOGY

Printed in Great Britain by
The Burlington Press (Cambridge) Ltd, Foxton, Cambridge

Published in 1987 by
Institute of Terrestrial Ecology
Merlewood Research Station
GRANGE-OVER-SANDS
Cumbria
LA11 6JU

BRITISH LIBRARY CATALOGUING-IN-PUBLICATION DATA
Angling and wildlife in fresh waters. —
(ITE symposium, ISSN 0263-8614; no. 19)
1. Freshwater ecology 2. Fishing —
Environmental aspects
I. Maitland, Peter S. II. Turner, Angela K.
III. Institute of Terrestrial Ecology
IV. Series
574.5'2632 QH541.5.F7
ISBN 0-904282-99-6

The *Institute of Terrestrial Ecology (ITE)* was established in 1973, from the former Nature Conservancy's research stations and staff, joined later by the Institute of Tree Biology and the Culture Centre of Algae and Protozoa. ITE contributes to, and draws upon, the collective knowledge of the 14 sister institutes which make up the *Natural Environment Research Council*, spanning all the environmental sciences.

The Institute studies the factors determining the structure, composition and processes of land and freshwater systems, and of individual plant and animal species. It is developing a sounder scientific basis for predicting and modelling environmental trends arising from natural or man-made change. The results of this research are available to those responsible for the protection, management and wise use of our natural resources.

One quarter of ITE's work is research commissioned by customers, such as the Department of Environment, the European Economic Community, the Nature Conservancy Council and the Overseas Development Administration. The remainder is fundamental research supported by NERC.

ITE's expertise is widely used by international organizations in overseas projects and programmes of research.

Dr P S Maitland
Institute of Terrestrial Ecology
Edinburgh Research Station
Bush Estate
PENICUIK
Midlothian
EH26 0QB

031 445 4343

Contents

Preface

Formerly regarded as a harmless healthy outdoor pursuit available to people from all walks of life in all parts of the world, angling has met with considerable criticism over recent years. The 'noble art' described by Walton in 1653 is now regarded by many as a cruel sport, often pointless (especially where the fish are returned to the water and not eaten) and somewhat alien in an increasingly conservation-conscious society. Much of the criticism of anglers in recent years has come from the increasing numbers of 'animal rights' pressure groups who have broadened their interests from concern over mammals and birds to cold-blooded vertebrates.

The situation has also been exacerbated by the increasing numbers of anglers themselves and the greater amount of leisure time (and unemployment) available to them. These developments have resulted in greater pressure on fish populations and the waters in which they live, and created concern among conservationists about increasing damage to wildlife caused by the various activities of anglers. Concern has been expressed about, among other matters, the direct depletion of fish stocks, the harmful effects of moving fish into new waters, and damage and distress caused by lead weights, nylon line and other materials left in the countryside.

One of the great difficulties in the debate over anglers and wildlife has been the lack of relevant facts on which to base any objective judgement. Anglers themselves are notoriously subjective in their approach to the sport and jokes concerning the supposed size of a catch or 'the one that got away' are traditional. In spite of the fact that anglers have been fishing in the British Isles for hundreds of years and now number over 3 million individuals, the lack of scientific data related to their activities and their impact on the environment is astonishing.

Because of all this recent interest in the subject, it was decided to hold an open meeting on the impact of anglers on wildlife in order to review the problems which were giving rise to concern and the related scientific data which might be available. At the suggestion of R W Edwards, the Scottish Freshwater Group and the British Ecological Society agreed to arrange such a meeting and a small organizing committee was set up by P S Maitland (SFG) and A K Turner (BES) to include the Institute of Fisheries Management (A V Holden) and the Nature Conservancy Council (D Bayne). Support and interest were also given by the Royal Society for the Protection of Birds.

It was decided to hold a one-day meeting as part of the regular series organized by the Scottish Freshwater Group at the University of Stirling, and local arrangements were made by D S McLusky. The Institute of Terrestrial Ecology agreed to publish the papers presented at the meeting as one of their regular series of symposia proceedings. The meeting was held on 30 October 1985, and 180 participants from a wide range of disciplines attended. The topic attracted some attention from the media, and press and television reports followed next day.

Peter S Maitland
Angela K Turner
June 1986

Lead poisoning and waterfowl

G J THOMAS*, C M PERRINS† and J SEARS†
*Royal Society for the Protection of Birds, Sandy, Bedfordshire
†Department of Zoology, University of Oxford, Oxford

1 Introduction

Although waterfowl are occasionally poisoned at natural lead deposits (eg old mines) and from organic pollution incidents involving alkyl lead and other compounds, the main cause of lead poisoning amongst waterfowl is from waste inorganic lead distributed in the environment by humans (Thomas 1980). Lead poisoning amongst waterfowl after ingesting waste lead cartridge pellets has been shown for 15 countries where research has been carried out. In Great Britain, mute swans (*Cygnus olor*) have been shown to die in large numbers after ingesting waste anglers' lead weights, and this problem is also known amongst the introduced mute swan populations in North America.

Lead poisoning affects the neuro-muscular systems. Quarry species suffering from lead poisoning are more prone to be shot. Large weight losses are recorded, and birds are likely to starve to death as a result of the gizzard muscles being unable to digest food which builds up in large compact masses in the crop.

2 Waterfowl and spent cartridge pellets

Spent lead pellets from cartridges are known to poison waterfowl in both Europe and North America (Belrose 1959; French & Baker 1982). Most research has been done in North America where 2–3% of the waterfowl populations overall are estimated to die each year of lead poisoning. In recent years, this probably means a loss of some 1.6–2.4 million birds. In the United Kingdom, it has been tentatively and conservatively estimated that at least 8000 mallard (*Anas platyrhynchos*) alone die from this cause between September and February each year (Mudge 1983).

Research in the United States and the United Kingdom has shown that spent lead pellets remain in the top few centimetres of aquatic substrate, while they settle down further in soft substrates. There is, however, a net accumulation of pellets in the surface layers. Pellet densities of 30 per square metre have been found in the top 10 cm of substrate at baited shooting ponds in Britain, and 16 per square metre at the internationally important waterfowl site at the Ouse Washes in Cambridgeshire. Densities of pellets at wetlands overshot by clay pigeon shooting clubs in Denmark have been measured at a staggering 20 000 per square metre.

The susceptibility of waterfowl to picking up spent lead pellets seems, in part at least, dependent on their feeding habits. Herbivores, such as wigeon (*Anas penelope*), feeding clear of the substrate seem to have a lower incidence of lead pellets in their gizzards. Ducks, such as pochards (*Aythya ferina*) and pintail (*Anas acuta*), which grub into the substrate for plant and invertebrate foods, have the highest incidences of pellets in their gizzards.

Waterfowl suffering from lead poisoning or wounded (and still containing lead pellets in their bodies) are more readily preyed upon than healthy birds and may give rise to cases of secondary poisoning. This has recently been shown to have been the case with bald eagles (*Haliaeetus leucocephalus*) in the United States, where deaths from lead poisoning are now a significant problem. Also, 3 of the last 4 dead Californian condors (*Vultur gryphus*) autopsied have died from lead poisoning after ingesting lead rifle bullets. It is suspected that they were ingested after the condors had fed upon wounded deer that had been shot by hunters.

The United States Government has progressively implemented a system of non-toxic (steel) shot zones in parts of some 40 States over the last 13 years to protect wildfowl populations. Additionally, the use of lead-shotted cartridges has been banned in many counties, usually following river valleys, to protect bald eagles and other waterfowl predators. The Danish Government has recently banned the shooting of lead pellets over water to protect waterfowl populations. In all cases, hunters are using recently developed and improved (second generation) steel-shotted cartridges. These perform just as effectively as the lead heads at all recommended shooting ranges. There is no difference in the crippling rates with these new cartridges and, after planned distribution, they are as competitive in price as the corresponding premium lead ones.

3 Mute swans and anglers' weights

An estimated 3370–4190 mute swans die in England each year from lead poisoning, with smaller numbers dying in Wales and Scotland. Bewick's swans (*Cygnus bewickii*) are also known to be affected (French & Baker 1982). Lead poisoning after ingesting discarded anglers' weights is the single biggest killer of mute swans. Ingested lead gunshot accounts for only 1% of the swans dying from lead poisoning.

About half of all mute swans examined by the Ministry of Agriculture, Fisheries and Food from a wide variety of locations in England have died from lead poisoning. This finding is about the same as the results from the swans examined along the River Thames. About 75% of mute swans examined from East Anglia died from lead poisoning (French 1984).

Mute swan numbers in Great Britain were estimated at 18 000 in 1983, which is some 5–13% below those estimated in 1955–56. This drop has occurred over a period of time when they should have increased, as has happened with some other species of waterfowl. The largest decreases have occurred at the major coarse fishing areas in lowland Britain. In contrast, increases are noticeable in those parts of Great Britain where fly fishing predominates and few lead weights are used.

Along the River Thames, swans that have died of lead poisoning have an average of 6 partially digested anglers' weights (mostly split shot) in their gizzards. In 1984, a particularly bad year, 66% of the Thames cygnets became ill, or died from lead poisoning before they were 10 weeks old. One 6-week-old cygnet contained 13 anglers' split shot.

Seasonal variation is high, with much higher incidences of deaths from lead poisoning occurring in the summer months. Blood lead levels of living swans mirror the pattern of these deaths. In 1983–85, 95% of swans on the lower River Thames had blood lead levels of over 40 μg/100 ml, which is taken as the threshold level indicating exposure to abnormally high levels of lead.

4 Discussion

The British Government has accepted that the lead poisoning of mute swans is serious and has promised legislation banning the use, sale and import of a range of lead weights from 1 January 1987. A wide range of non-toxic weights has been produced by manufacturers over the last 3 years, and these have gained acceptance with the main angling organizations (Thomas 1985).

The voluntary measures adopted in 1984 had mixed success, as shown by a survey of tackle shops carried out by the Royal Society for the Protection of Birds. Lead shot sold in spill-proof containers was available at almost all tackle shops, but lead continued to be sold in non-spill-proof containers at most shops. Point of sale publicity (posters and codes of conduct) warning of the dangers of lead weights to swans and other wildlife was evident at less than half of the tackle shops visited. Alternatives to lead weights were available at over half of the tackle shops visited in major coarse fishing areas, but were not being purchased by anglers. A survey of anglers carried out by the Anglian Water Authority (French 1984) reported that only 3% were using alternative weights.

5 Summary

Waterfowl in many countries are poisoned by lead from human sources. The 2 main offenders are waterfowl hunters and anglers. Lead in cartridges can be replaced by steel and in anglers' weights by various satisfactory compounds. The onus lies with the 2 groups of people involved to use these substitutes and minimize further damage to wildlife.

6 References

Bellrose, F. C. 1959. Lead poisoning as a mortality factor in waterfowl populations. *Bull. Ill. St. nat. Hist. Surv.*, **27**, 235–288.

French, M. C. 1984. Lead poisoning in mute swans – an East Anglian survey. In: *Metals in animals*, edited by D. Osborn, 25–29. (ITE Symposium no. 12.) Abbots Ripton: Institute of Terrestrial Ecology.

French, M. C. & Baker, L. 1982. Lead hits Bewick's swans. *BTO News*, **121**, 1.

Mudge, G. P. 1983. The incidence and significance of ingested lead pellet poisoning in British wildfowl. *Biol. Conserv.*, **27**, 333–372.

Thomas, G. J. 1980. A review of ingested lead poisoning in wildfowl. *Bull. int. Waterfowl Res. Bur.*, **46,** 43–60.

Thomas, G. J. 1985. Lead poisoning in swans. *RSPB Information Leafl.*, 1985, 1–2.

Angler litter

R W EDWARDS and M CRYER
Department of Applied Biology, University of Wales Institute of Science and Technology, Cardiff

1 Introduction

It seems remarkable that, despite the damage caused to some species of aquatic birds by lead weights and other forms of angler litter, there have been so few descriptions of the distribution of litter at sites, or of those angling activities and site characteristics relevant to the problem. Several factors contribute to this omission. First, the methods of search are generally tedious and labour intensive, and may be inefficient or applicable only to parts of the habitat. Second, recorded litter items need not necessarily be available, or a danger to wildlife. Furthermore, the conversion of measures of site contamination into past rates of litter addition relative to angler use is frequently difficult to undertake, so restricting the utility of the information for management purposes.

Table 1 classifies freshwater angling activities in relation to their potential generation of litter, specific and non-specific to anglers. Lead is associated with coarse fishing, where split shot and ledgers are widely used, but is also used to a small extent in various forms of fly and bait game fishing. The discarding of line results particularly from entanglement during the casting of flies and floats, and from the replacement of terminal tackles. Hooks, rarely abundant in angler litter, are most easily lost when attached to line in the more active forms of fishing. It is widely recognized that, in fly fishing, such hooks are frequently snagged on bankside obstructions and vegetation, whereas in float fishing most are snagged underwater or on marginal vegetation. Float fishermen commonly use, and may discard, small fine-wired hooks which become blunt or bent in use. General litter left by anglers is particularly prevalent at coarse angling sites for a number of reasons: the location of individual anglers is generally fixed during a visit; fishing sessions are often long; and angling practices may involve the use of polythene groundbait and 'caster' bags, and tin cans for sweetcorn and meat baits, which may be discarded when empty (Forbes 1986).

Table 1. Litter problems and their association with different forms of angling. The number of asterisks indicates the potential severity of the problem as assessed by the authors

Form of angling		Lead						
		Shot	Ledger	Nymph	Line	Hooks	Bait	General
Game	Fly	–	–	*	***	**	–	*
	Spin	–	–	–	*	*	–	*
	Bait	*	*	–	**	*	*	*
Coarse	Float	***	–	–	**	**	***	***
	Ledger	–	**	–	*	*	***	***

In addition to these categories of 'inadvertent' litter, groundbaiting is commonly an adjunct to coarse angling in particular. Cereal groundbait and maggots are frequently left in large quantities at fishing sites – generally, but not exclusively, below the water line (M Cryer *et al.*, unpublished data). Apart from its general ecological effects on waterbodies (Edwards & Fouracre 1983), such bait may provide a food source for riparian birds and mammals, and in so doing act as an attractant to areas containing high densities of potentially more damaging litter.

This paper reviews the few studies at sites in Great Britain where quantitative information is readily avail-

Table 2. Details of sites surveyed in Wales

Site and location	Area (ha)	Fishing bank (m)	Species caught *	No. of pegs	Angler use ** (d/year)	Age as fishery (year)	Other uses ***
Hendre (S Glamorgan)	4	920	C,R,B,	52	1000	6	C
Llanishen (Cardiff)	21	1900	Tr	–	15500	>10	S,O
Llandrindod (Powys)	7	1000	C,R,B	58	4000	>10	B
Roath Park (Cardiff)	11	1200	C,R	open access	9700	>10	B
Tredegar Park (Gwent)	4	600	C,R	30	10600	>10	B
Woodstock (Gwent)	1	460	C,R,T	55	5700	6	O

* C = carp (*Cyprinus carpio*); R = roach (*Rutilus rutilus*); B = bream (*Abramis brama*); T = tench (*Tinca tinca*); Tr = trout (*Salmo* sp.)
** At some sites, estimates of angler use are very approximate
*** C = canoeing; S = sailing; O = bird-watching; B = boating

able (Table 2), and draws attention to future requirements for data on angling litter in relation to wildlife protection and environmental management. Unfortunately, sites for which data are available are all still waters, in Wales. This introduces a systematic bias to our review, as conditions on flowing water would be expected to be different, especially in other parts of Great Britain where angling pressures may vary.

2 Methods

Several methods have been used to assess the quantity of lead at fishing sites. Visual, on-site searches are only practicable on hard, smooth substrates, such as tarmac, concrete, or impacted earth. More generally, samples of bank and shallow-water substrate must be removed and searched using sieving or drying techniques in the laboratory. Radiographs of substrate material have facilitated the identification of lead and other metal objects in one study (J Sears, personal communication). Metal detectors have been used both on the banks and in shallow water (Plate 1), but their

Plate 2. Laying a plastic sheet in front of a fishing 'peg' to determine the rate of deposition of lead shot and other angler litter (Photograph D V Bell)

Plate 1. Metal detectors have been shown to be biased and inefficient when used to determine the distribution of lead shot (Photograph D V Bell)

Plate 3. Visual bankside searches: the only practical method of retrieving discarded nylon monofilament. Llanishen Reservoir, Cardiff (Photograph D V Bell)

efficiency is generally low and dependent on substrate type and target size. Membranes, covered with uncontaminated substrate material, have been placed in shallow water in front of fishing pegs (Plate 2), and accumulation of lead in the time between setting and retrieval has been related to angler use at the site (Bell *et al.* 1985).

Unless attached to line, hooks are not easily found, except by careful searches of the substrate. In consequence, their abundance in the environment is likely to be underestimated.

Line is recovered using visual search methods (Plate 3), but these are only effective on land. The low specific gravity and 'facies' of discarded line probably result in considerable transport by wind and water currents, leading to possible bias of discard rates if they are periodically estimated by local bank searches.

General litter is principally a 'site-amenity' problem, although some categories of litter, such as plastic bags and tin cans, present a hazard to certain animals. The amount of litter has rarely been assessed, but 2 measures that have been used are: the number of classified litter items per unit of bank frontage (M Cryer *et al.*, unpublished data); and their weight and perceived area (Forbes 1986).

Almost all these methods have been devised to determine what litter is present at sites, rather than what is accessible to potentially vulnerable wildlife. However, the lake periphery at fishing sites, where most litter is deposited, is both the focus of the majority of surveys, and the feeding zone for several bird and mammal species. More detailed studies of the behaviour of given taxa would be required before effective sampling procedures to assess litter availability could be developed. Furthermore, it seems likely that combinations of litter items (such as lead shot or hooks attached to line), and interactions between litter and other site factors (such as water depth and bank profiles, or vegetation) are of considerable significance to the vulnerability of wildlife. In past surveys, the importance of these interactions has not always been recognized.

3 Results

3.1 Lead shot

Where fishing positions or pegs are identifiable at sites, the bankside and shallow-water distribution of lead shot is clearly associated with these positions. Forbes (1986) calculated that 95% of all the lead shot recovered from Llandrindod Lake (Table 2) was within 2.5 m of angling pegs, most being in front of them, rather than behind.

Recent initial investigations at 2 sites in south Wales, Hendre and Tredegar Park (Table 2), have confirmed that densities of lead shot are highest within 2.5 m of angling positions. Whilst there is considerable small-scale patchiness in the distribution of shot in this zone of high deposition, there is a tendency for concentrations to occur in front of, and to the right of, angling positions (Figure 1). This concentration probably results from the conscious discarding of shot by predominantly right-handed anglers.

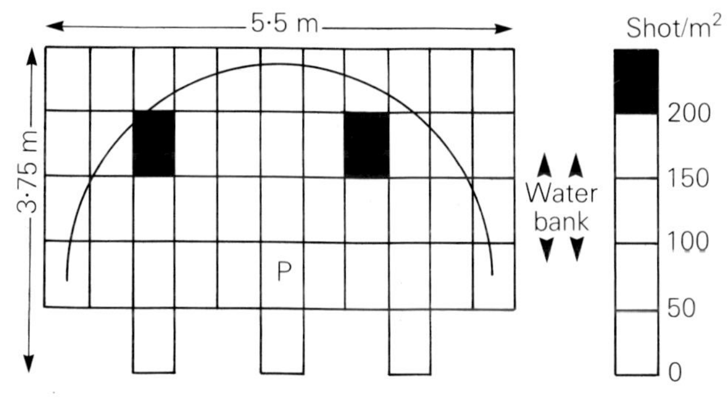

Figure 1. Mean densities of lead shot recovered from samples taken on a standard grid from 3 sites at Hendre Lake, Cardiff. An arc of radius 2.5 m from the angler position 'P' is shown

During 1984, Bell *et al.* (1985) surveyed the distribution of lead shot in front of fishing pegs at Woodstock, a site specifically developed for coarse fishing (Table 2). Average densities of shot decreased rapidly with distance from the angler position, being about 230 shot/m^2 at the pond margin, and 60 shot/m^2 only 2 m from the shore. By laying membranes in front of fishing pegs during the first 80 days of the fishing season, and relating lead shot accumulation to angler visits, it was estimated that anglers discarded, on average, 2–3 lead shot per visit – equivalent to about 5 kg for the fishery in one season. At this level of shot deposition, the initial quantities of lead recovered near fishing pegs represented far less than a single season's loss.

At Roath Park, where fishing positions are not confined to pegs, but are continuous along the shore, an initial survey consisting of 25 shallow water transects, normal to the shore, has similarly shown the pronounced decrease in the density of lead shot with distance – in this case, from about 330 shot/m^2 at the bankside, to around 60 shot/m^2 at 2.25 m from the shore. Figure 2 shows the fall-off in density of shot in relation to distance from the shore for the 4 sites in south Wales.

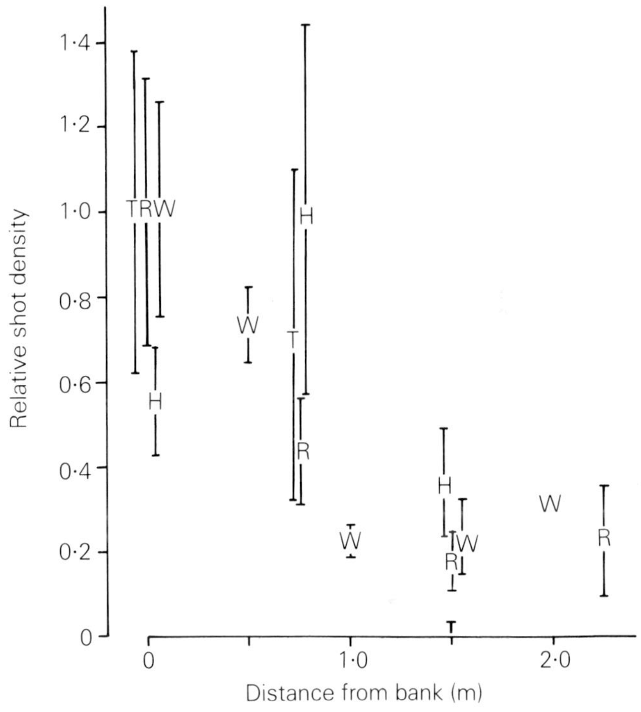

Figure 2. Relative mean densities of split shot (+SE) with increasing distance from the shore at 4 still-water sites in south Wales. In each case, relative mean density is calculated as absolute mean density divided by maximum mean density for that site (R = Roath Park, with maximum density of 327 shot/m^2; W = Woodstock, with maximum density of 230 shot/m^2; T = Tredegar Park, with maximum density of 89 shot/m^2; H = Hendre, with maximum density of 80 shot/m^2)

Studying the general distributions of lead shot around Llandrindod Lake, Forbes (1986) estimated that at least 33.5 kg of lead shot was present on the lake shore and fishing platforms, 99.7% of it being present in the 18% of the lake shore clearly identified as normally used for fishing.

The weights of recovered lead shot have been determined at all 5 still-water sites studied, and size distributions are very similar at all of them (Table 3). The most abundant size is nearly always BB (0.4 g), which on average comprises 20% of all shot recovered.

Table 3. Percentage distribution of lead shot among manufacturers' standard size categories at 5 still-water sites in Wales. Overall mean distributions by number and weight are also shown

	Percentage of shot in standard size categories (along with weight in g)										
Site (no. of shot examined)	SSG 1.89	AAA 0.81	BB 0.40	#1 0.28	#2 0.24	#3 0.20	#4 0.17	#5 0.13	#6 0.10	#7 0.08	<#8 <#0.06
Llandrindod (N = 1404)	5.2	11.2	17.0	16.9	6.1	15.0	8.6	4.7	7.5	6.1	1.8
Woodstock (N = 399)	2.0	10.8	21.7	15.4	–	8.5	12.1	10.6	8.2	3.4	7.2
Hendre (N = 126)	2.4	11.8	23.6	12.6	–	11.8	13.4	8.7	6.3	0.8	8.7
Tredegar (N = 51)	–	3.9	19.6	11.8	–	15.7	13.7	7.8	11.8	5.9	9.8
Roath (N = 96)	3.1	10.4	19.8	10.4	1.0	16.7	11.5	6.3	5.2	2.1	13.5
Average by number (5 sites)	2.5	9.6	20.3	13.4	1.4	13.5	11.9	7.6	7.8	3.7	8.2
Average by weight (5 sites)	18.6	24.3	25.4	11.7	1.1	8.4	6.3	3.1	2.4	0.9	1.5

Several sizes, eg 2, 5, and 7, are not normally sold as split shot for fishing, so accounting for their complete absence at some sites, and their scarcity at others. The proportionate weights of lead contributed by the different size categories, as an average value for all sites, are also shown in Table 3, and are compared with proportionate numbers, as a cumulative plot, in Figure 3. Although size categories 1, BB, AAA, and SSG contribute 76% of the lead burden, they constitute only 46% of the number of recovered shot.

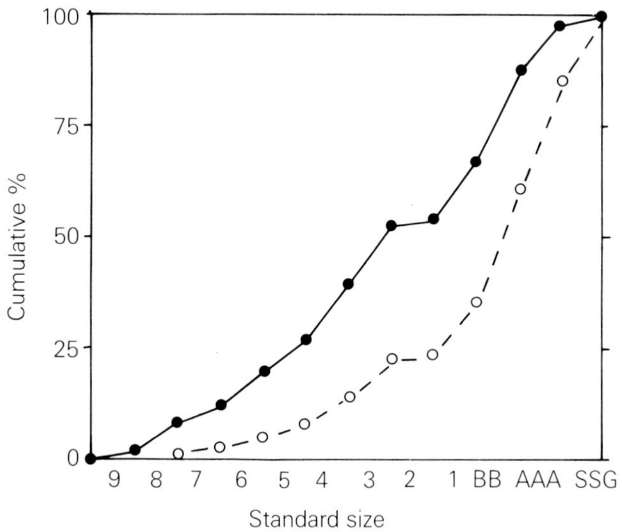

Figure 3. Cumulative percentage plot by number (●), and by weight (O) of increasing sizes of split lead shot recovered from 5 still-water sites in Wales

Ledger weights have been omitted from these comparisons, and, whilst they were rarely found at most sites, at Llandrindod Lake they comprised 2.6% of the lead weights recovered, and therefore contributed substantially (50%) to the total lead burden.

In relation to the accessibility of lead to wildlife, it may be important to discriminate between loose lead shot, and that attached to line. At Roath Park, 10.4% was attached to line; at Hendre, 3.2%; and at Tredegar Park, 3.9%. At Llandrindod, 13% of recovered lead was attached to line, half of this being on relatively short pieces of nylon. At Woodstock, there was a marked contrast in the proportions of recovered lead attached to line in the 2 sampled areas: 100% in marginal vegetation and only 9% on sediments recovered from submerged membranes (Bell *et al.* 1985). This contrast may well be an artefact of the different sampling methods employed in the 2 areas.

The depth of core and turf samples removed from sites may be critical in any assessment of the amount of lead available to wildlife, because it is possible that only shot in the upper sediment layers is accessible. Studies to date have determined total shot density only, and any future survey of available lead should include data on vertical as well as horizontal distribution.

3.2 Line

From Table 1, it will be noted that substantial amounts of discarded line may be associated with both coarse and game angling, particularly where float or fly fishing is prevalent. Bell *et al.* (1985) collected line from 50-m lengths of bank at intervals throughout the fishing season at Llanishen Reservoir, a fly-only trout fishery where reasonable estimates of angler visits could be made. In initial collections, an average of 7.8 m of nylon monofilament line per metre of bank was recovered. Subsequent searches throughout the season indicated that this accumulation represented line discarded over about 20 weeks of fishing. Anglers, on average, left between 2.5 m and 5 m of line per visit.

At Woodstock, during the 6 years of its operation as a coarse fishery, an average of about 75 m of line had accumulated within the marginal and shallow water area of each fishing peg – equivalent to about 4 km at the whole site or 9 m per metre of site perimeter. Subsequent studies using membranes recorded only underwater deposition rates, and these were variable but low – equivalent to about 0.2 m per angling visit. Whilst it may be tempting to compare deposition rates at the game and coarse fisheries, the 2 sites were sampled very differently, no periodic underwater surveys being conducted at Llanishen, and no periodic bank searches at Woodstock.

At the coarse fishery at Llandrindod Lake (Forbes 1986), where searches were confined to the shore and marginal areas, 50 m of nylon line per metre of bank were recovered, most occurring at very high densities in fishing areas, which comprised only 18% of the shoreline. It was estimated that about 57 km of line had accumulated on the banks of the fishery. Size distributions of line fragments recovered from Llandrindod and Llanishen are compared in Figure 4.

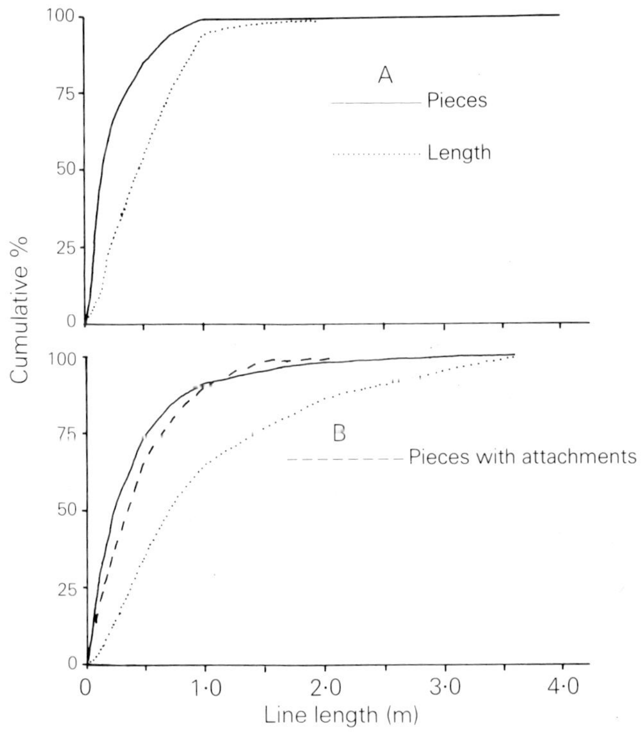

Figure 4. Cumulative distributions of pieces, and total lengths of line within increasing length categories at A, Llanishen Reservoir (fly fishing only), and B, Llandrindod Lake (coarse fishing only). In B, the distribution of line pieces with attached hooks or lead shot is also shown

The accessibilty of line to wildlife will depend, in part, on the lengths of individual pieces, and on their location. Furthermore, damage after encounter or ingestion will be strongly influenced by attachments, either shot or hooks. Information on some of these properties of discarded line is available at a few sites. At Llandrindod, line pieces with and without attachments were analysed separately (approximately 12% of pieces had attachments), but distributional characteristics were very similar. At Woodstock, Bell *et al.* (1985) noted that 31% of line pieces had hooks or shot attached, and that size distributions were similar for both categories – mean lengths being 0.30 m and 0.23 m respectively.

Table 4 gives percentile values for distributions of line lengths, and of total lengths within classes, for Llandrindod and Woodstock (coarse sites), and Llanishen (fly-only game site). Fifty and 90 percentile values of line lengths were less than 0.35 m and 1 m respectively for all sites. Inevitably, with respect to total length of line recovered in each of the length categories, small individual pieces represent a much smaller proportion of the whole, 50% of the total length of recovered line from all sites being represented by pieces greater than 0.38 m in length.

Table 4. Percentage number of pieces of line (A) and total length of line (B) in relation to piece length at 3 still-water fisheries in Wales. Maximum lengths for each percentile are given in metres

	Percentage of total number of pieces		
A	20	50	90
Llandrindod	0.09	0.25	1.00
" with attachments	0.10	0.35	1.00
Llanishen	0.08	0.15	0.70
Woodstock	0.06	0.13	0.52
	Percentage of total length recovered		
B	20	50	90
Llandrindod	0.30	0.70	2.50
Llanishen	0.20	0.45	0.90
Woodstock	0.15	0.38	0.83

3.3 Hooks

Detached hooks in samples are probably underestimated, particularly the small ones commonly used in coarse fishing. Nevertheless, at Woodstock, hooks were 3% as abundant in litter as lead shot. In all surveys of coarse fisheries, most recovered hooks are attached to line (Bell *et al.* 1985; Forbes 1986), a form in which they may present a greater hazard to wildlife. At the one game fishery that has been surveyed (Bell *et al.* 1985), bank searches returned 19 flies and hooks – equivalent to one fly per 20 m of line, or perhaps 4–8 angler visits.

3.4 General litter

Only one detailed survey of general litter at a fishery has been undertaken (Forbes 1986), and, being a lake in the spa town of Llandrindod Wells, collections of litter were made relatively frequently. Although the site was used by visitors other than anglers, 64% of the number of litter items (93% of the total surface area of litter) were recorded in those parts of the shoreline (18%) predominantly used by anglers.

An island at the site, used exclusively by anglers, was particularly affected by litter. The highest litter densities for the whole site were found around the fishing platforms on the island, but other areas throughout the island were also badly contaminated, with a high proportion of items (48%) being discarded bait containers. The problems of litter on this island were exacerbated by the difficulty of refuse collection from the area. The major contribution of anglers to the litter problem at Llandrindod was reinforced by ordination analysis, which identified the litter-strewn sites as those with high use by anglers and low use by general visitors.

3.5 Groundbait

Despite the reluctance of many water engineers to permit groundbaiting in reservoirs because of potential problems with water quality, the practices of cereal- and loose-feeding are widespread on most coarse fisheries. The contributions made by groundbait to the organic carbon and phosphorus budgets of reservoirs have been assessed by Edwards and Fouracre (1983), and in most cases these represent less than 1%, and less than 0.2% of the total budget respectively. Whilst this particular effect of groundbaiting has been overemphasized in the past, no study seems to have been made of the role which groundbait plays, either as a food source for wildlife, particularly during the winter, or in attracting mammals and birds to areas where the density of lead shot, hooks, or other litter is high.

4 Discussion

It is unfortunate that detailed information on litter distribution is available at so few fishing sites, most of which are still waters in Wales. Comparisons with and between flowing waters are particularly constrained, data being restricted to the River Thames (J Sears, personal communication). There is an evident need to extend the range of sites, relating measurements of litter distribution to wildlife use and angler behaviour. Whilst sampling methodologies need to be standardized and rigorously tested to facilitate site comparisons for both bank and underwater areas, it is also important that descriptions of litter distribution are relevant in relation to wildlife vulnerability. In particular, the absence of information both about the vertical distribution of lead shot in underwater sediments and the feeding horizons of vulnerable wildlife species restricts the usefulness of the gross litter densities that are available.

Furthermore, the rate of deposition of litter needs to be determined in relation to existing accumulations so that the responsiveness of environmental improvements to changing management and fishing practices, as well as the time period over which wildlife is potentially vulnerable, can be assessed. The few studies that have included measurements of deposition rate have concluded that such rates are high when compared with accumulations. Bell *et al.* (1985) calculated a deposition rate at Woodstock Pool of 2–3 split shot per day using a membrane technique. Comparing this to densities before membrane laying, it was estimated that the total site burden represented only 4–5 months' accumulation – a rather surprising conclusion in view of the sustained use of the site for the previous 6 years. Unless there is substantial 'emigration' of shot from sampled near-shore areas, one must assume far more rapid degradation rates of lead than are reported in the literature (at least 25 years; Thomas 1982), provided both litter accumulations and deposition rates are being accurately measured. The resolution of these issues is important in assessing the period over which wildlife will remain vulnerable to lead ingestion following the withdrawal of this material from use. The precise vertical location of lead shot in sediment, particularly as it affects the redox potential to which shot is exposed, could be crucially important in determining rates of solution by both chemical and microbial actions.

There is a consistency in the spatial distribution of lead shot at the predominantly still-water sites surveyed. All studies identify a zone of high deposition around angling positions which is remarkably localized. Mean densities of lead shot at Welsh sites around angler locations ranged from 50 to 200/m^2 (after allowing for a sampling efficiency varying between 60% and 94% with shot size). On the Thames, densities varied from 60 to 580/m^2, being highest at Richmond where lead poisoning of mute swans (*Cygnus olor*) was most severe (J Sears, personal communication). Comparable data on effects on wildlife at Welsh sites are not available.

Vulnerability of wildlife, whilst depending in part on the precise location of lead shot, may also be affected by its attachment to line. At Welsh sites, between 3% and 13%, and in the Thames 11%, of shot were attached to line (Bell *et al.* 1985; Forbes 1986; J Sears, personal communication). Hooks similarly pose a greater threat when attached to line, although the number of hooks recovered in surveys is generally low, perhaps as a result of the fairly rapid degradation of the small fine-wired hooks frequently used in freshwater angling. The role of groundbait in attracting wildlife to 'high litter' areas needs urgent assessment.

Discarded line has been a persistent problem at fishing sites in the past: the RSPCA reported dealing with the entanglement of 1500 birds in 1974 (C Booty, personal communication). As with shot and hooks, the hazard from line probably becomes greater when combined with other forms of angler litter. Bell *et al.* (1985) recorded 23% of pieces of recovered line at Woodstock, and Forbes (1986) noted 11% at Llandrindod Lake having shot or hook attachments. Fluorescent line, introduced to increase line visibility for reasons of tackle control, has been suggested as an alternative to traditional monofilament to facilitate both avoidance by wildlife and collection by anglers. Its effectiveness in either objective has not yet been adequately tested, but at a recent fly fishing competition in Wales, the capture rate for trout with fluorescent line was less than 10% of that with 'normal' line. Clearly, further developments are required before an acceptable angling substitute could be recommended. It is possible that the general use of fluorescent running lines with traditional end-tackles might provide an effective compromise, although there is some evidence that end-tackle is both more hazardous and more readily discarded, so limiting the usefulness of combined arrangements.

It has been reported that incidents involving injury to wildlife from angler litter show distinct seasonal trends. During 1981–82, a survey by RSPCA inspectors of swan injuries (C Booty, personal communication) indicated a marked peak from June to August when angling activity was most intense, and an extremely low incidence during April and May when no coarse angling occurred. Similarly, J Sears (personal communication) recorded marked seasonality of lead concentrations in the blood of swans on the Thames, where fishing took place, but no seasonality at sites where fishing was banned. However, such peaks occurred in February and September, and it is not yet clear whether they are related solely to angling activity and the availability of recently deposited lead or to seasonal changes in feeding behaviour and food sources at this location. If the availability of recently discarded lead rather than seasonality in feeding behaviour is the important factor determining hazard, a rapid decrease in blood lead levels and mortality might be expected with the introduction of lead substitutes. If, however, discarded lead remains accessible to wildlife for long periods of time, the localized nature of its deposition around angling locations may facilitate its practical removal.

General litter has usually been viewed as a site amenity problem (Plate 4) (Pardee *et al.* 1981), although anglers may, for several reasons, be major contributors (Forbes, 1986). Some categories of litter – such as plastic bags and tin cans – although not exclusive to anglers, may be discarded by them after being used as bait containers, and so create hazards to wildlife. More attention needs to be given to the management of fishing sites and the 'architecture' of angling locations to minimize litter deposition and to facilitate litter collection. Apart from general zoning of sites for different recreational activities, etc, the specific location and design of angling positions, with the provision of litter bins and impervious surfaces for easy litter recovery, might prove worthwhile where site usage is high or damage to wildlife causes concern. Such damage cannot be regarded solely as a conservation issue and confined to angling impact at the population level; it is legitimately one of animal welfare, too.

Plate 4. A site amenity problem: litter recovered from Hendre Lake after just 2 weeks of the new fishing season in 1985 (Photograph M Cryer)

5 Summary

Information on the distribution of angler-specific and general litter at 5 still-water game and coarse fisheries in Wales is reviewed. Lead shot is concentrated around angler locations, 95% at one site being within a radius of 2.5 m. At 4 other sites, the concentration on the underwater substrate 1.5 m from the shore was only 20–20% of that at the water's edge (up to 330 shot/m^2); 2–3 lead shot per angler visit were discarded at one site. The most common size at all sites was BB (0.4 g) which accounted for 20% of all recovered shot. Excluding ledgers, the 3 heaviest categories (SSG, AAA, BB) represented 76% of all lead, by weight. Between 3.2% and 13% of recovered lead shot were attached to line. Between 8 m and 50 m of line per metre of bank had accumulated at 3 contrasting fisheries: most pieces (90%) were more than 1 m long, and at one site 30% had attached hooks or shot. At a game fishery, anglers left, on average, 2.5–5.0 m of nylon monofilament line per visit. General litter was surveyed at one coarse fishery: despite being used by many visitors, most litter was concentrated around angler locations, and included containers used for bait. The relevance of litter distribution to wildlife protection and site management is discussed. More detailed studies of wildlife and angler behaviour are needed to assess and minimize hazards of litter. Further data are required on current patterns of litter accumulation in relation to site use and management.

6 Acknowledgements

Financial assistance for this work was provided by the Manpower Services Commission; Welsh Water Authority; and UWIST. We are grateful to the site managers at Tredegar Park, Roath Park, and Hendre Lake for permission to work on their lakes. Finally, thanks are due to the other members of the group, Russell Bull,

Justin Corbett, Kim Favager, Nigel Linley, Greg Whittle, and Mike Winterbotham, without whose toils this analysis would not have been possible.

7 References

Bell, D. V., Odin, N. & Torres, E. 1985. Accumulation of angling litter at game and coarse fisheries in south Wales, UK. *Biol. Conserv.*, **34**, 369–379.

Edwards, R. W. & Fouracre, V. A. 1983. Is the banning of groundbaiting in reservoirs justified? *Proc. Br. Freshwat. Fish. Conf., 3rd, Liverpool, 1983,* 89–94.

Forbes, I. J. 1986. The quantity of lead shot, nylon fishing line and other litter discarded at a coarse fishing lake. *Biol. Conserv.*, **38**, 21–34.

Pardee, L., Tarbet, D. D. & Gregory, J. D. 1981. Management practices related to recreational use and the responses of recreationists to those practices along the French Broad River in North Carolina. *Proc. Warmwater Streams Symp., Tennessee, 1980,* 21–30. American Fisheries Society (Southern Division).

Thomas, G. J. 1982. Lead poisoning in waterfowl. In: *Managing wetlands and their birds*, edited by D. A. Scott, 260–268. (Publication no. 14.) International Waterfowl Research Bureau.

Disturbance by anglers of birds at Grafham Water

A S COOKE
Nature Conservancy Council, Northminster House, Peterborough

1 Introduction

In Britain, there have been a number of local studies on the effects on waterfowl of recreational activities, including fishing (Hume 1976; Batten 1977; Tydeman 1977a; Tuite *et al.* 1983; Watmough 1983; Kramer 1984; UWIST 1984; Bell & Austin 1985). In addition, broad overviews of the subject have been undertaken by Tanner (1979) and Tuite *et al.* (1984).

Grafham Water in Cambridgeshire is one of the most important wintering sites for waterfowl in Britain (Tanner 1979). The main recreational pursuits are sailing and game fishing. The former activity has relatively little impact on waterfowl on this very large reservoir, because it occurs only intermittently and disturbance is more or less confined to deep water which tends to be avoided by the majority of waterfowl (Cooke 1977). On the other hand, bank and boat fishermen arrive in large numbers on every day during the fishing season, and often fish the shallow, sheltered bays and creeks that are favoured by the birds. Consequently, fishermen have a considerable effect on waterfowl at Grafham (Cooke 1976, 1977). Since these earlier accounts were published, studies on disturbance at Grafham have continued. All observations relating to fishing disturbance are brought together in this paper, the aims of which are as follows.

i. To determine the extent to which fishing affects numbers of the principal species of water birds at Grafham.
ii. To determine the extent to which fishing affects their distribution between fished and unfished areas.
iii. To determine how and why the reactions of mallard (*Anas platyrhynchos*) to a standard shore-based disturbance are modified in areas disturbed by the public and by fishermen. In this respect, the abundant mallard is used as a convenient indicator of the types of reactions that might be shown by other species.
iv. To quantify the reactions of water bird species to a standard shore-based disturbance outside the fishing season in order to determine whether such observations might help explain the changes in numbers and distribution noted during the fishing season.
v. To document, for as many species as possible, their relative reactions to disturbance, in the hope that this may be of relevance for decision-making at other sites facing similar environmental pressures.

2 The site

Grafham Water was created in the mid-1960s by building a dam across a shallow river valley and pumping in water from the River Great Ouse. A map of the reservoir is shown in Figure 1. At top water, Grafham is about 700 ha in size and, when first created, was the largest reservoir in England.

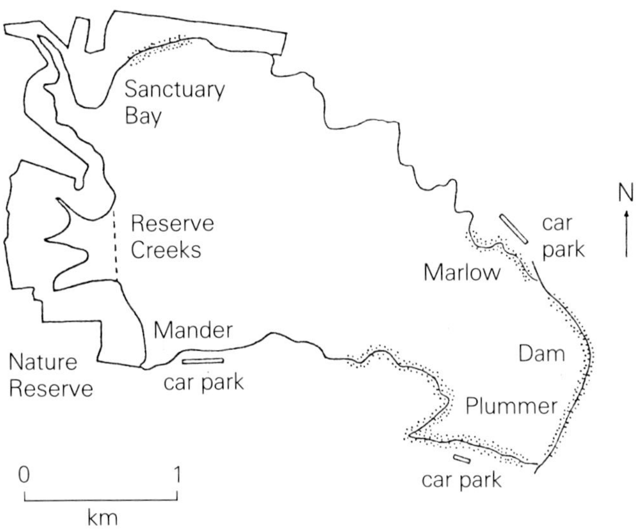

Figure 1. A map of Grafham Water showing the features referred to in the text. Shoreline study areas for the controlled approaches are indicated by stippling

Various formal and informal recreational activities are available. Game fishing and sailing both result in a considerable human presence on and around the water. Diurnal and seasonal changes in these 2 activities were discussed by Cooke (1977). During the early 1970s, the trout fishing season extended from the beginning of April to mid-October. The end of the season was put back to the end of October during the early 1980s. In 1985, the season was extended till the end of December for bank fishermen. During the early 1980s, there have been, on average, roughly 80–100 rods per day, with 20–30 motor boats per day being hired (Anglian Water Authority statistics). The principal unfished area is the Nature Reserve Creeks (Figure 1); these creeks represent 2–3% of the water area of the reservoir.

Grafham also fills the niche of the local park; there is parking space for 1400 cars and the picnic areas cover about 40 ha. The public has direct access to about half of the shore. The Nature Reserve comprises the western and north-western shores of the lake (Figure 1). Most of its 170 ha is arable land which acts as a buffer zone and helps to keep the central section of the reserve comparatively peaceful.

Direct observations on the approachability of water birds were undertaken to fulfil aims (iii) and (iv). These observations involved walking towards birds and noting their reactions (see Section 3), and were carried out mainly in 3 study areas at the eastern end of the lake (Figure 1). Birds in these areas are subject to different amounts of shore-based disturbance by the public (Table 1). However, disturbance by fishermen and other water-based amenity users tends to be similar in the 3 areas. It is a reasonable assumption that there is more land-based disturbance from the public at Marlow than at either Plummer or the Dam, but it is not possible to state whether there is more disturbance at Plummer than at the Dam because the nature of disturbance is not the same in the 2 areas. It should be noted, however, that, compared with lightly disturbed areas in the north-west of the reservoir, even Plummer and the Dam suffer from considerable disturbance from humans.

Table 1. Details of the areas used for the controlled approach studies. See methods for explanation

Area	Capacity of car park	% of shore to which public allowed	% of shore to which fishermen allowed
Plummer	200 cars	40	100
Marlow	700 cars	100	100
Dam	No car park	0 but allowed along the length of the road behind the Dam wall	100

3 Methods

Since the reservoir first began to flood in January 1965, water birds at Grafham have been counted regularly for the Wildfowl Trust. Grebes (*Podiceps* and *Tachybaptus* spp.), wildfowl and coot (*Fulica atra*) (here collectively referred to as waterfowl), waders, grey heron (*Ardea cinerea*), cormorants (*Phalacrocorax carbo*), divers (*Gavia* spp.), etc, have been counted on the middle weekend of each month from September through to March. Usually 2 observers have counted in a systematic, synchronized fashion, one counting from the north shore, working west to east, and the other operating along the south shore. This technique was adopted to minimize the risk of missing birds or of counting them more than once. The reservoir was demarcated into 6 areas (one of these, the Reserve Creeks, is shown in Figure 1), and totals for each area have been recorded separately. This series of counts has been augmented by many other counts, both of the whole reservoir and of particular count areas, undertaken at different seasons of the year. During the period 1972–75, many additional counts were made to facilitate more precise understanding of the redistribution that occurs at the end of each fishing season (Cooke 1976, 1977). These counts have been re-examined in this paper, especially to answer objective (i).

From 1979 to 1985, direct observations were made on birds' reactions in the eastern study areas. Individuals or groups were selected on land or on the water, but within 10 m of the water's edge. I recorded the number of birds and the nature of their activity, and then walked towards them at a standard rate (see Cooke 1980). For an individual, the distance at which it moved off was recorded (termed the tolerance distance), as was the nature of the movement and how far the bird moved. For a group, the median reaction was recorded. Tolerance distances were estimated by pacing. The mean length of 10 paces was 10.05 ±SE 0.04 m (N=10), so one pace was taken to be equivalent to 1 m. Distances moved were estimated by eye up to 100 m. Longer distances were simply recorded as 'greater than 100 m'. When judging distances out on to the water, advantage was taken of any reference point a known distance away. I tested my ability to judge distance in this manner by estimating distances from 10 m to 100 m on an open grass field, and then estimating by pacing. There was no tendency for the visual estimate to be greater or smaller; the mean difference for 28 observations was +2.8 m. Ten visual estimates of 100 m had, when paced out, a mean ±SE of 104±7 m.

4 Results and observations

4.1 The effect of fishing on numbers and distribution

Counts summarized in Table 2 show the mean number of waterfowl from 5 weeks before fishing ended, 1972–75, until 5 weeks after. During this part of the autumn, waterfowl numbers increase (Cooke 1977), and a general upward trend is evident in Table 2. However, the only significant change between adjacent periods was the rise (of 28%) at the end of the fishing season.

Table 2. Changes in numbers of Grafham's waterfowl (wildfowl, grebes and coot) relative to the end of fishing, 1972–75

Period relative to end of fishing season (days)	No. of counts	Mean no. of waterfowl ± SE
−34 to −14	8	3830 ± 400
−13 to −7	5	4820 ± 420
−6 to 0	5	4440 ± 420
+1 to +7	5	5680 ± 290*
+8 to +14	3	5530 ± 380
+15 to +35	11	5350 ± 390

*Significantly greater than mean for previous period, $P<0.05$

Changes in the numbers of individual species at the end of the fishing seasons, 1972–75, are presented in Table 3 for waterfowl, waders and other water birds. Although only 2 species showed significant increases associated with the end of fishing, the means increased for 11 out of the 14 species listed in Table 3, a trend which was statistically significant ($\chi^2=4.57$, $P<0.05$). The relative increase during the first days of the close season may indicate the extent to which species avoided Grafham immediately prior to the end of the fishing season.

Table 3. Mean numbers of birds at Grafham during the last week of fishing (5 counts) and the first week of the close season (5 counts), 1972–75. List includes all species for which the mean count was at least 5 for at least one period

Species	Fishing Mean ± SE	No fishing Mean ± SE	% change in means
Grey heron	1.2 ± 0.4	5.2 ± 1.7*	+333
Pochard	89 ± 42	211 ± 33	+138
Wigeon	27 ± 9	45 ± 21	+65
Tufted duck	698 ± 182	1006 ± 84	+56
Little grebe	15 ± 4	22 ± 8	+47
Mallard	844 ± 59	1200 ± 88**	+42
Gadwall	31 ± 10	43 ± 8	+39
Great crested grebe	62 ± 17	71 ± 16	+15
Teal	268 ± 60	309 ± 47	+15
Mute swan	85 ± 13	98 ± 11	+15
Coot	2220 ± 296	2530 ± 249	+14
Shoveler	83 ± 14	62 ± 26	−26
Dunlin	94 ± 33	66 ± 18	−30
Ringed plover	5.0 ± 1.0	3.4 ± 2.2	−32

* Significantly greater than mean for fishing, P<0.05; ** P<0.01
Latin species names given in Table 14

The effect of fishing on total numbers is, however, much less striking than the effect on distribution. The extent to which species gathered in the unfished Reserve Creeks is used here to indicate which species were most affected. Data from September and October counts made during the fishing seasons, 1972–84, are presented in Table 4. In the same Table, comparable data for the non-fishing period, from the end of October to December, are included to indicate which species tend to be found in the Creeks outside the fishing season. For 8 of the 14 species listed in Table 4, on average more than half the birds were recorded in the Reserve Creeks during the fishing season. For 7 species, numbers in the Reserve Creeks decreased significantly after fishing ended, despite total numbers at Grafham being maintained or elevated: gadwall (*Anas strepera*), coot, grey heron, mallard, great crested grebe (*Podiceps cristatus*), tufted duck (*Aythya fuligula*) and little grebe (*Tachybaptus ruficollis*). Numbers of pochard (*Aythya ferina*) in the Reserve Creeks decreased by about 50%, although total numbers increased significantly. Outside the fishing season, the teal (*Anas crecca*) showed the greatest attraction for the Reserve Creeks, followed by dunlin (*Calidris alpina*), grey heron, mallard and gadwall (Table 4).

4.2 The tolerance distance of mallard in specific areas in relation to disturbance

During the fishing season, the mean tolerance distance of mallard was significantly lower (ie the birds were tamer) in the severely disturbed Marlow area, than in the 2 moderately disturbed areas (Table 5). After fishing finished, there were significant increases in mean tolerance distance in each of the areas (Table 5). Birds remained significantly more approachable at Marlow than at Plummer or the Dam. However, it should be stressed that the last 2 areas are not

Table 4. Mean numbers of birds in the Reserve Creeks and in total for Grafham, during fishing, September–October, 1972–84 (26 counts). List includes all species recorded on more than half of the counts. Comparable data for the close season are also given, October–December, 1972–84 (41 counts)

Species	Fishing Mean ± SE Reserve	Fishing Mean ± SE Total	Mean percentage in reserve	No fishing Mean ± SE Reserve	No fishing Mean ± SE Total	Mean percentage in reserve
Pochard	36 ± 12	47 ± 13	77	18 ± 5	158 ± 19	11
Gadwall	18 ± 4	24 ± 5	75	4.8 ± 1.8**	25 ± 4	20
Teal	212 ± 33	290 ± 40	73	118 ± 34	306 ± 49	38
Coot	1050 ± 110	1520 ± 160	69	281 ± 66***	1850 ± 150	15
Grey heron	2.0 ± 0.4	3.1 ± 0.6	64	0.8 ± 0.2*	3.2 ± 0.4	25
Mallard	730 ± 70	1170 ± 80	62	263 ± 32***	1170 ± 70	23
Great crested grebe	38 ± 6	67 ± 6	56	7.8 ± 1.1***	147 ± 19	5
Shoveler	21 ± 5	41 ± 6	51	5.6 ± 2.1***	34 ± 4	17
Wigeon	14 ± 3	31 ± 14	44	23 ± 10	199 ± 39	12
Tufted duck	216 ± 36	543 ± 58	40	64 ± 15***	876 ± 95	7
Little grebe	2.7 ± 0.8	7.3 ± 1.4	37	0.8 ± 0.2**	6.3 ± 1.2	12
Dunlin	3.9 ± 1.5	31 ± 9	13	14 ± 4	53 ± 11	26
Ringed plover	0.2 ± 0.1	6.7 ± 1.6	4	0.02	0.2 ± 0.1	14
Mute swan	3.0 ± 0.8	92 ± 9	3	2.8 ± 0.7	44 ± 7	6

* Significantly less than mean for fishing, P<0.05; ** P<0.01; *** P<0.001

Table 5. Mean tolerance distance of mallard in the study areas during the fishing season and the close season, 1979–85

Area	Fishing Level of disturbance	Fishing No. of groups	Fishing Mean tolerance distance ±SE (m)	No fishing Level of disturbance	No fishing No. of groups	No fishing Mean tolerance distance ±SE (m)
Plummer	Moderate + fishing	16	89 ± 12**	Moderate	61	141 ± 8††***
Dam	Moderate + fishing	11	76 ± 11*	Moderate	79	179 ± 9†††***
Marlow	Severe + fishing	12	43 ± 5	Severe	82	66 ± 4†

* Significantly different from mean for Marlow, P<0.05; ** P<0.01; *** P<0.001
† Significantly different from mean for fishing, P<0.05; †† P<0.01; †††P<0.001

especially quiet, and the birds there showed some tolerance of humans. Thus, the mean tolerance distance of 13 observations in the lightly disturbed area of Sanctuary Bay (see Figure 1) was 324 m (SE±27 m). This mean was significantly greater than that at either Plummer or the Dam ($P<0.001$).

To understand why changes in tolerance distance occurred at the end of the fishing season, the situation has been examined in detail at the Dam, the study area where the change was most marked (Table 5). During the fishing season, birds that reacted at distances greater than 200 m were absent (Table 6). Explanations for this observation include (i) wilder birds avoided the Dam during fishing, and (ii) the same birds were present throughout, but reacted at a lesser distance during fishing. Data collected for the autumn periods of 1982 and 1983 indicate the former explanation to be the more likely (Table 7). Mallard numbers on the Dam increased as soon as fishing ended; at the same time, tolerance distance increased significantly. Waterfowl move out of the Reserve Creeks as soon as fishing ends (Cooke 1976, 1977), and it appears that wild birds at Grafham are more or less confined to the Reserve Creeks during fishing, but then quickly occupy other areas when fishing finishes, and so modify the observed mean tolerance distance in those areas. It was considered possible that the observed differences at areas such as the Dam might be due, at least in part, to birds being wilder because group size increased (eg Owens 1977). However, when data were examined for the Dam, it was found that, outside the fishing season, the increase in tolerance distance was independent of group size (Table 8). This finding again supports the theory that the observed change in tolerance distance at the eastern end when fishing ends is due to the arrival of wilder birds from the western end (and perhaps other waters).

Table 6. Tolerance distances of groups of mallard in the Dam area, 1979–85, comparing fishing and close season data

Range of tolerance distances (m)	Number of groups: Fishing	No fishing
1–49	3	2
50–99	6	10
100–149	1	17
150–199	1	16
200–249	–	17
250–299	–	10
300–349	–	6
350–399	–	–
400+	–	1
Total	11	79

Table 7. Changes in the mean tolerance distance and numbers of mallard in the Dam area at the end of fishing, 1982–83

Period	No. of groups	Mean tolerance distance ± SE (m)	No. of visits	Mean no. of birds (range)
Fishing, August–October	9	64 ± 8	8	7 (0–29)†
No fishing, October[×]	6	137 ± 23*	1	73
No fishing, November	23	175 ± 14**	6	37 (2–107)
No fishing, December	18	186 ± 14**	5	30 (0–71)

[×] One day after fishing ended, 1983
* Significantly greater than mean for fishing/August–October, $P<0.01$, ** $P<0.001$
† Significantly less than combined no fishing/October–December, $U_{8,12}=21.5$, $P<0.05$

Table 8. Relationship between group size and mean tolerance distance for mallard in the Dam area, 1979–85

	Fishing		No fishing	
Group size	No. of groups	Mean tolerance distance ± SE (m)	No. of groups	Mean tolerance distance ± SE (m)
1–9	10	77 ± 12	50	185 ± 12*
10–99	1	58	29	168 ± 11

* Significantly different from mean for fishing, $P<0.001$

4.3 The comparative approachability of water birds

To try to predict which species might be vulnerable to disturbance from shore-based activities, the reactions of different species have been compared outside the fishing season. All species that were recorded at least 10 times at the eastern end are listed in Table 9, in increasing order of tameness. The goldeneye (*Bucephala clangula*) and those species higher in the ranking order were significantly less approachable than the mallard ($P<0.05$), while the redshank (*Tringa totanus*) and those below it were significantly tamer than the

Table 9. Mean tolerance distance towards an approaching human for species at the eastern end of Grafham during the close season 1979–85. List includes all species recorded on at least 10 occasions

Species	No. of groups	Mean tolerance distance ± SE (m)
Gadwall	47	181 ± 12
Grey heron	32	178 ± 13
Coot	87	169 ± 7
Goldeneye	29	168 ± 14
Great crested grebe	10	142 ± 22
Tufted duck	12	131 ± 18
Mallard	222	127 ± 5
Shelduck	19	126 ± 17
Shoveler	28	126 ± 15
Wigeon	73	115 ± 6
Black-headed gull	12	105 ± 10
Redshank	19	92 ± 8
Teal	56	86 ± 6
Ringed plover	37	43 ± 3
Pied wagtail	10	32 ± 6
Dunlin	27	30 ± 4
Meadow pipit	11	27 ± 2
Mute swan	18	14 ± 10

mallard (P<0.05). The data are expressed in Table 10 in a manner that may be more useful to planners. This Table indicates the degree of sensitivity of each species to an approach to a specified distance.

Table 10. The extent of movement by species when approached on foot to specified distances, eastern end, close season data, 1979–85. Numbers of observations are given in Table 9

	% of groups moving when approached to specified distance						
Species	350m	300m	250m	200m	150m	100m	50m
Great crested grebe	0	0	10	30	40	50	60
Grey heron	3	9	16	34	56	91	100
Mute swan	0	0	0	0	6	6	6
Shelduck	0	0	5	11	26	74	95
Wigeon	0	0	3	5	25	60	96
Gadwall	4	11	21	40	62	79	96
Teal	0	0	0	4	11	38	84
Mallard	1	4	9	20	36	57	87
Shoveler	0	4	7	21	32	46	93
Tufted duck	0	0	0	27	55	64	100
Goldeneye	3	3	3	17	38	52	79	100
Coot	0	5	10	30	54	86	100
Ringed plover	0	0	0	0	0	0	30
Dunlin	0	0	0	0	0	0	15
Redshank	0	0	0	0	5	37	89
Black-headed gull	0	0	0	0	8	42	100
Meadow pipit	0	0	0	0	0	0	0
Pied wagtail	0	0	0	0	0	0	20

When considering the likely impact of disturbance on birds, it is, however, not sufficient simply to consider the distance at which birds react; how they react should also be taken into account. Species that tend to fly when disturbed are more likely to be displaced greater distances. For instance, in Table 10, where comparative data are given for gadwall and wigeon (*Anas penelope*), it is apparent that any approach down to 100 m is more likely to elicit a move in a flock of gadwall than in a flock of wigeon. However, gadwall tend to swim away from such disturbance, whereas wigeon fly away, and, as shown in Table 11, an approach within 200 m will cause greater numbers of wigeon to move more than 100 m.

Table 11 shows the extent to which each species moved more than 100 m when approached to 300, 200, 100 or 0 m. Not surprisingly, there was a significant positive relationship for the 18 species between the percentage of groups that flew when approached to 0 m and the percentage moving more than 100 m (Spearman rank correlation coefficient=0.714, P<0.01). Few birds swam more than 100 m. In effect, there are 4 ranking orders in Table 11, each indicating the likely impact on exposure to disturbance of varying intensity. Which of these orders is most applicable to Grafham can be arrived at by ruling out unrealistic ranking orders one by one. Thus, distribution and numbers of dunlin and ringed plover (*Charadrius hiaticula*) are unaffected by fishing at Grafham (Tables 3 & 4), so the 'approach to 0 m'

Table 11. The extent of movement of more than 100 m when approached on foot to specified distances, eastern end, close season data, 1979–85. Numbers of observations are given in Table 12

	% of groups moving >100 m when approached to specified distance				% of groups that flew when approached
Species	300m	200m	100m	0m	to 0m
Great crested grebe	0	10	10	20	0
Grey heron	9	31	84	94	100
Mute swan	0	0	0	0	0
Shelduck	0	5	16	26	37
Wigeon	0	4	38	59	93
Gadwall	2	4	13	24	28
Teal	0	0	31	82	95
Mallard	0	2	6	13	40
Shoveler	4	7	21	50	68
Tufted duck	0	0	0	9	9
Goldeneye	3	28	52	66	62
Coot	1	4	5	5	10
Ringed plover	0	0	0	54	100
Dunlin	0	0	0	52	100
Redshank	0	0	28	89	100
Black-headed gull	0	0	25	83	100
Meadow pipit	0	0	0	9	100
Pied wagtail	0	0	0	0	100

ranking order seems to represent disturbance that is too severe. Approaching to 200 m (or 300 m) caused virtually no impact, except to grey heron or goldeneye. On the other hand, approaching to 100 m produces no long-distance movement in tamer species but appreciable responses in wilder species, and this order is given in full in Table 12 as perhaps representing a situation relevant to Grafham.

Table 12. Ranking order showing percentage of groups moving more than 100 m when approached to 100 m, eastern end, close season data, 1979–85

Species	Number of groups	Number moving >100 m	Percentage
Grey heron	32	27	84
Goldeneye	29	15	52
Wigeon	73	28	38
Teal	55	17	31
Redshank	18	5	28
Black-headed gull	12	3	25
Shoveler	28	6	21
Shelduck	19	3	16
Gadwall	46	6	13
Great crested grebe	10	1	10
Mallard	218	14	6
Coot	85	4	5
Tufted duck	11	0	
Mute swan	18	0	
Dunlin	27	0	0
Ringed plover	35	0	
Meadow pipit	11	0	
Pied wagtail	10	0	

4.4 A comparison of ranking orders

In this paper, 4 different types of ranking order have been presented. There are 2 empirical orders based on the extent to which birds came to Grafham after

Table 13. A comparison of the ranking orders for the 12 species common to each order

	Derived from bird counts		Derived from controlled approaches	
	I Immigration when fishing ended	II Preference for Reserve during fishing	III Mean tolerance distance	IV Tendency to move >100 m when approached to 100 m
Most wild	Grey heron	Gadwall	Gadwall	Grey heron
	Wigeon	Teal	Grey heron	Wigeon
	Tufted duck	Coot	Coot	Teal
	Mallard	Grey heron	Great crested grebe	Shoveler
	Gadwall	Mallard	Tufted duck	Gadwall
	Great crested grebe }	Great crested grebe	Mallard	Great crested grebe
	Teal }	Shoveler	Shoveler	Mallard
	Mute swan }	Wigeon	Wigeon	Coot
	Coot	Tufted duck	Teal	Tufted duck }
	Shoveler	Dunlin	Ringed plover	Mute swan }
	Dunlin	Ringed plover	Dunlin	Dunlin }
Most tame	Ringed plover	Mute swan	Mute swan	Ringed plover }

Spearman rank correlation coefficients:

	I	II
III	0.483 (NS)	0.734 (P<0.01)
IV	0.531 (P<0.05)	0.633 (P<0.01)

fishing (Table 3) or preferred the Reserve Creeks during fishing (Table 4). Then, there are 2 predictive orders, based on how birds reacted when approached by an observer (Tables 9 & 12). Twelve species appear in all of these Tables, and these are listed in order in Table 13. Ranking tests between the empirical-predictive pairs showed that the best fits occurred (i) between the preference for the reserve order and the tolerance distance order, and (ii) between the immigration order and the order showing tendency to move more than 100 m. In addition, there was a significant relationship between the preference for the reserve order and the tendency to move more than 100 m order.

5 Discussion

Fishing has been reported as seemingly affecting numbers of teal, pochard and coot at Llangorse Lake (Tuite *et al.* 1983), pochard, shoveler (*Anas clypeata*), wigeon, mallard and tufted duck at Foremark (Watmough 1983) and coot at Llanishen (UWIST 1984). Mallard and tufted duck at Llangorse (Tuite *et al.* 1983) and coot at Foremark (Watmough 1983) were unaffected. Distributional changes, similar to those caused by fishing at Grafham, have also been noted at Rutland Water (Appleton 1982). Rutland is even larger than Grafham, and its extensive Reserve supports most of the birds during the fishing season. Changes in distribution caused by fishing have also been observed for coot, wigeon, mallard and pochard at Llandegfedd (UWIST 1984; Bell & Austin 1985) and for mallard, pochard, tufted duck and coot at Roath Park Lake (UWIST 1984). Each species that was mentioned as being tolerant of disturbance was found to be affected elsewhere, and one is left with the conclusion that any species is likely to be affected at a badly disturbed locality. Tuite *et al.* (1984) examined count data for 9 wildfowl species for the winter months from hundreds of sites, and determined whether those sites with certain recreational activities had significantly lower numbers of birds. Coarse fishing was found to be the most damaging activity, but game fishing was reported as having no effect, perhaps because there is no game fishing for most of the winter. In contrast to this conclusion, Watmough (1983) found that duck tended to avoid sites in the north Midlands that had intensive trout fishing. Tuite *et al.* (1984) ranked their 9 study species according to the frequency with which they displayed decreases associated with recreation. The goldeneye was found to be most susceptible, followed by the shoveler, teal, wigeon and mallard. The remaining species (pochard, tufted duck, goosander (*Mergus merganser*) and mute swan (*Cygnus olor*) showed no such decreases. Of these, the pochard is the most surprising, especially since Tuite *et al.* (1983) found marked effects of recreation on numbers of pochard at Llangorse Lake.

At Grafham, fishing has some effect on bird numbers and a marked effect on distribution for many species. Disturbed birds are presumably more likely to redistribute rather than leave at a large water such as this, because they may find somewhere quiet to settle. The importance of the Reserve Creeks as a sanctuary area has been demonstrated in this paper. If fishing were allowed in the Reserve Creeks, this would cause an additional reduction of about 60% in waterfowl numbers if the Reserve's birds vacated Grafham. When planning a nature reserve at a reservoir used for recreation, it is important to ensure that it is large enough and/or sufficiently screened so as to minimize the effects of disturbance. At Grafham, fishing boats are allowed up to the mouth of the Reserve Creeks, a position in which they are very conspicuous. Most of the waterfowl congregate at the back of the Creeks, 200–400 m away from the nearest fishing boats. A smaller reserve would presumably have been less successful in maintaining Grafham's waterfowl population during the fishing season (see Watmough 1983). Over the last 10 years, willow growth (*Salix* spp.) on the shore, especially in the Plummer area, has proved attractive to birds (and unattractive to fishermen). Numbers of mallard have decreased in the Reserve and increased in the Plummer area since the mid-1970s (Cooke & Mason 1983).

The distribution of birds at Grafham is influenced by other human activities, as well as fishing. This influence is evident from the fact that, even during the fishing season, approachability of mallard and other species (unpublished observations) at the eastern end was related to the degree of public pressure. It is reasonable to suggest that any increase in disturbance will further modify distribution and numbers. A development of concern in this respect was the decision to hold a water-skiing trial in the autumn of 1985. The sudden increase in the length of the fishing season in 1985 is also of concern. During November 1985, there appeared to be few bank fishermen interested in continuing their sport. If, however, boats are made available during future winters and the extension gains in popularity, bird numbers and distribution will presumably be modified further.

Few data on approachability have been published. Hume (1976), Tydeman (1977b) and Batten (1977) reported distances to which birds allowed approach by boats of different types, but observations were limited in number. Watmough (1983) reported the distance to which waterfowl species could be approached on foot: mallard and tufted duck were significantly less approachable than coot. This situation differed from that at Grafham, where coot were significantly wilder than mallard (Table 9). Indeed, coot were less approachable than mallard in each of the 3 study areas at Grafham (unpublished observations).

Owens (1977) found that for Brent geese (*Branta bernicla*) on the Norfolk coast, tolerance distance increased as flock size increased. No such relationship existed for mallard observed in the Dam area of Grafham (Table 8), but extending the study across the 3 study areas outside the fishing season would have produced a positive trend, as smaller groups of mallard occurred in the Marlow area where birds were tamer.

Until detailed observations are undertaken elsewhere, it is not possible to state how applicable the ranking lists derived for Grafham might be for other waters. Information on the susceptibility of birds to leave or redistribute at Grafham is summarized for all species in Table 14. Although at most other waters there will be less opportunity for disturbed birds to find a quiet sanctuary area within the water itself, such conditions might exist close by. Thus, coot disturbed at Llanishen Reservoir simply moved across a causeway to the adjacent and quieter Lisvane Reservoir (UWIST 1984). At Grafham, birds disturbed at the Dam end may move more than 3 km to quiet areas, such as the Reserve Creeks.

Table 14. Overall ranking orders derived from bird counts and controlled approach studies

	Susceptibility to redistribution	Tendency to be absent
Grey heron (*Ardea cinerea*)	++	++
Pochard (*Aythya ferina*)	++	++
Goldeneye (*Bucephala clangula*)	++/+	++
Gadwall (*Anas strepera*)	++	+
Coot (*Fulica atra*)	++	o
Little grebe (*Tachybaptus ruficollis*)	+	+
Wigeon (*Anas penelope*)	+	+
Teal (*Anas crecca*)	+	+
Mallard (*Anas platyrhynchos*)	+	+
Tufted duck (*Aythya fuligula*)	+	+
Redshank (*Tringa totanus*)	+/o	+/o
Black-headed gull (*Larus ridibundus*)	+/o	+/o
Great crested grebe (*Podiceps cristatus*)	+	o
Shelduck (*Tadorna tadorna*)	+	o
Shoveler (*Anas clypeata*)	+	o
Mute swan (*Cygnus olor*)	o	o
Ringed plover (*Charadrius hiaticula*)	o	o
Dunlin (*Calidris alpina*)	o	o
Meadow pipit (*Anthus pratensis*)	o	o
Pied wagtail (*Motacilla alba*)	o	o

++Susceptibility/tendency is marked; + apparent; o not apparent

At other waters faced with the threat of a new disturbing operation involving human activity on the shore, it may be worth testing the reactions of mallard to indicate the contemporary level of disturbance. If wintering mallard typically tolerate controlled approaches on foot (as used at Grafham) to within 100 m, then there is already severe disturbance. However, if the typical reaction is to move away at distances of 300 m or more, then this will indicate birds that are unaccustomed to much disturbance, and therefore presumably susceptible to any increase in disturbance.

At Grafham, the impacts of (i) combined water- and shore-based activity and (ii) a standard shore-based activity have been studied experimentally. Although there was good agreement between observations from the 2 studies, Watmough (1983) has indicated that birds on the water are more susceptible to water-based activity, and *vice versa*. Thus, the more aquatic species such as pochard, goldeneye, tufted duck and grebes may be relatively more susceptible to water-based disturbance than is indicated in Tables 10 and 12. Bearing in mind the ranking order in Table 14, pochard and goldeneye are likely to be amongst the species most affected. Other local studies have consistently noted effects on these species (Hume 1976; Tuite *et al.* 1983; UWIST 1984; Bell & Austin 1985).

At the other end of the scale, waders were relatively unaffected, although redshank were significantly less approachable than ringed plover or dunlin (Table 9), and were more prone to fly further than 100 m (Table 11). These results are not surprising as the wary nature of the redshank is well known amongst bird-watchers. The resistance to disturbance of the mute swan and the small passerine species was also to be expected (eg see Cooke 1980).

While one should never lose sight of conservation implications on a broader regional or national scale, management decisions concerning recreation and birds are usually made at a local level. As Watmough (1983) has argued, for scientific site management, one needs to know whether disturbance matters to the birds, by interpreting information on the response of birds and the time taken to recover in terms of their biology. At Grafham, there has been some impact of fishing on bird numbers. While it is not possible to state categorically whether this impact matters to the birds involved, it should be appreciated that the missing birds are likely to be the more wary members of Grafham's winter waterfowl community, and therefore the least likely to find suitable quiet conditions nearby. It is also worth pointing out that the missing birds render Grafham less attractive to bird-watchers. Conversely, however, the uneven distribution caused by fishing is helpful to bird-watchers by facilitating viewing from the hide overlooking the Reserve Creeks. Again, it is not possible to answer whether having to occupy a different area of the lake has a detrimental effect on the individuals involved. Even after fishing has finished, however, bird distribution remains modified by public pressure. Provision of an adequate refuge area would seem essential on any lake where recreation activities are commonplace, if that site is to retain a bird population consistent with its geographical position and physical and chemical attributes.

6 Summary

It was first noticed in the early 1970s that trout fishermen at Grafham Water, Cambridgeshire, markedly affected the distribution of waterfowl, with most birds being found in the relatively small, unfished area of the Nature Reserve Creeks. This paper brings together observations made at the reservoir throughout the 1970s and early 1980s. For most duck species and for the coot, grey heron and great crested grebe, more than 50% of the birds occurred in the Reserve Creeks during fishing; but for the mute swan and waders, less than 15% were recorded there. Associated with the end of the fishing season, there was an average influx of waterfowl to Grafham of 28%. Most duck species increased by more than 30% within a week of the end of fishing, but numbers of coot, great crested grebe, mute swan and waders changed little. Despite this influx to Grafham, numbers of most species decreased in the Reserve Creeks. Controlled approaches on foot were made towards mallard to determine their tolerance to this form of disturbance: at the end of the fishing season, there was a decrease in tolerance in each of 3 study areas. It was concluded that, during fishing, comparatively few mallard were present, but these birds were highly tolerant; during the first few days of the close season, less tolerant birds arrived from quieter areas. Controlled approaches were also made to a range of other species during the close season. Significant positive relationships were found between the wariness of species and both the degree to which the Reserve Creeks were used during fishing and the extent to which species increased after fishing finished. The species most susceptible were grey heron, pochard, goldeneye and gadwall.

7 Acknowledgements

I am grateful to K D Mason for help with the bird counts throughout the period covered by the paper, and to Dr L Batten for helpful comments on the manuscript.

8 References

Appleton, T. 1982. Rutland Water Nature Reserve: concept, design and management. *Hydrobiologia*, **88**, 211–224.

Batten, L. A. 1977. Sailing on reservoirs and its effects on water birds. *Biol. Conserv.*, **11**, 49–58.

Bell, D. V. & Austin, L. W. 1985. The game-fishing season and its effects on overwintering wildfowl. *Biol. Conserv.*, **33**, 65–80.

Cooke, A. S. 1976. The effects of fishing on waterfowl at Grafham Water. *Rep. Camb. Bird Club, 48th, 1974*, 40–46.

Cooke, A. S. 1977. *The birds of Grafham Water.* Huntingdon: Huntingdon Local Publications Group.

Cooke, A. S. 1980. Observations on how close certain passerine species will tolerate an approaching human in rural and suburban areas. *Biol. Conserv.*, **18**, 85–88.

Cooke, A. S. & Mason, K. D. 1983. Recent changes in waterfowl distribution at Grafham Water associated with growth of lakeside willows. *Huntingdon Peterb. Bird Rep. 1981*, 50–54.

Hume, R. A. 1976. Reactions of goldeneye to boating. *Br. Birds*, **69**, 178–179.

Kramer, D. 1984. The effects of recreational activities on the winter wildfowl population at Priory Park Lake, Bedford, during the winter of 1982–83. *J. Beds Hunts Nat. Trust 1983/84*, 34–46.

Owens, N. W. 1977. Responses of wintering Brent geese to human disturbance. *Wildfowl*, **28**, 5–14.

Tanner, M. F. 1979. *Wildfowl, reservoirs, recreation.* (Research report 5.) London: Water Space Amenity Commission.

Tuite, C. H., Owen, M. & Paynter, D. 1983. Interaction between wildfowl and recreation at Llangorse Lake and Talybont Reservoir, south Wales. *Wildfowl*, **34**, 48–63.

Tuite, C. H., Hanson, P. R. & Owen, M. 1984. Some ecological factors affecting winter wildfowl distribution on inland waters in England and Wales, and the influence of water-based recreation. *J. appl. Ecol.*, **21**, 41–62.

Tydeman, C. F. 1977a. The importance of the close fishing season to breeding bird communities. *J. environ. Manage.*, 5, 289–296.

Tydeman, C. F. 1977b. *Gravel pits as conservation areas for breeding bird communities.* Ph.D. thesis, Bedford College, University of London.

University of Wales Institute of Science and Technology. 1984. *The impact of anglers on wildlife and site amenity.* Cardiff: UWIST.

Watmough, B. 1983. *The effects on wildfowl of recreation at reservoirs in the mid-Trent valley.* Birmingham: Severn-Trent Water Authority.

The effects of angling interests on otters, with particular reference to disturbance

D J JEFFERIES
Nature Conservancy Council, Northminster House, Peterborough

1 Introduction

In terms of numbers, anglers form an important part of the bankside community using fresh waters in Britain. Angling and netting, either for sport or for food, have been practised by humans over many centuries. Some interaction with another fish predator, the otter (*Lutra lutra*), either directly or indirectly, seems inevitable. What is, or has been, the importance of this interaction in the past and in the present and future as anglers increase in numbers? The otter population of Britain has seriously declined and it now has legal protection. Also, being attractive, it has achieved some importance as a kind of 'figurehead' within the conservation movement.

In the past, attempted extermination has been the problem. At present, disturbance seems to be the most likely important interaction, at first examination. The otter is apparently very shy and largely nocturnal in the south. It is seldom seen, even by people who earn their livelihood from rivers. Increasing disturbance was listed among the likely problems for the declining otter population when the Joint Otter Group (O'Connor *et al.* 1977) reviewed the situation in 1977. However, since then, conservationists have become less concerned over this factor as more information has been gained. The present paper collects together the available, but scattered, data plus further unpublished observations on the subject in order to reach some objective conclusions.

2 Changes in the population of the otter and connections with fishing interests

2.1 Eighteenth and nineteenth centuries: the effect of keepering

Howes (1976), writing about Yorkshire, Nottinghamshire and Derbyshire, suggested that the otter population density was relatively high until at least the mid-18th century. The record for hunting of 9 otters killed in one day by one hunt in 1796 (Bell 1837) supports this suggestion. However, the piscivorous otter was considered to be in direct competition with people. Freshwater fish were important as a source of food and, later, sport. Consequently, a tradition of persecution arose with the aim of eradicating the species. With the improvement of the sporting gun in the late 18th and early 19th centuries and the gin trap, the means for eradication became available. The prosperity following the industrial revolution increased the numbers of the landowning society with the leisure time and the interest in shooting and fishing and who employed large numbers of gamekeepers to preserve their sporting quarry from predators. For example, 263 otters were killed on the Duke of Sutherland's estate alone from 1831 to 1834 (Nethersole-Thompson 1951). Descriptions in the various Victorian County Histories show that persecution of the otter by 'keepers occurred in most counties. By the end of the 19th century, the otter population must have been artificially low. Corresponding declines of the polecat (*Mustela putorius*), wild cat (*Felis silvestris*) and pine marten (*Martes martes*) populations occurred at the same time, ending with relict populations by 1900. An analysis by Langley and Yalden (1977) again points to the cause being persecution by gamekeepers, rather than habitat change. Unlike these species, as far as is known, the otter never became extinct in any county, perhaps because of the difficulty of trapping. The otter will not come to bait as it only rarely scavenges (Cuthbert 1973).

2.2 Twentieth century: recovery

An analysis of the hunting records of an old established otter hunt, the Culmstock, was undertaken by Chanin and Jefferies (1978). The number of 'finds' per 100 days hunting over the period covered by the data showed a regular increase from 61 in 1907 to 92 in 1955. This significant increase in hunting success sustained over a period of nearly 50 years suggests an improving population. Langley and Yalden (1977) describe similar recoveries in the populations of the polecat, wild cat and pine marten from the period of the 1914–18 world war, with recolonization of old territory. They related this pattern to the decrease in persecution by 'keepers. This occurred due to the reduction in 'keeper numbers during 2 world wars and the changing economic climate (Potts 1980). It seems probable that this factor is also the reason for the increase in otters in the Culmstock hunt country (Devon and Somerset) and possibly elsewhere.

2.3 The recent decline

Although Stephens (1957) was able to say of most River Board areas that they were 'well stocked' with otters, by the time her report had been published another and more severe decline in the British otter population had started. Again, an analysis (Chanin & Jefferies 1978) of hunting records showed that 11 out of 12 packs of hounds had experienced a sudden and serious decline in hunting success in the mid-1950s. The date of the start of the decline was established at 1957–58 and it occurred simultaneously over the whole country. It resulted in a marked change in distribution, with otters now sparse to absent over most of the Midlands and a fragmented population elsewhere in England and Wales (Crawford *et al.* 1979; Lenton *et al.* 1980). To have a suddenly occurring, simultaneous and severe population decline over most of the country, but most severe in the south and east, suggested a new environmental factor, particularly as it happened to an apparently improving population. Chanin and Jefferies (1978) examined the possible causes. In considering the disturbance caused by angling, figures for the numbers of salmon and trout fishing licences issued in Devon and Cornwall from 1950 to 1971 were obtained and showed a steady increase from 1950 in both counties, reaching treble the numbers by the late-1960s. However, there was no sudden increase or change in rate in 1957–58, and it is doubtful whether any vertebrate population would show such a sudden and dramatic consequence to a steady increase in disturbance when it reached a particular level. The cause fitting the known information best was water pollution, following the use of the toxic and persistent organochlorine insecticides as cereal seed dressings in 1956. Thus, although largely responsible for the earlier decrease, it is unlikely that angling interests played a significant part in the recent and present decline.

3 Disturbance

3.1 Reactions to anglers

The effects of disturbance by anglers can be ascertained with certainty only when an otter can be observed at the time. Such reported occurrences are few. Known examples, from which it is possible to build up a picture of the otters' reaction to various types of disturbance, are given below. Several were obtained the ideal way, by observing the behaviour of an otter that is being tracked by means of a radio-transmitter fitted to a leather harness (Mitchell-Jones *et al.* 1984), when it meets a particular type of disturbance (Plate 6). These were obtained during the joint research funded by the Nature Conservancy Council, Vincent Wildlife Trust, Otter Trust and World Wildlife Fund, and undertaken with (i) J and R Green, tracking females P1, P2 and male P3 on the River Earn and its tributaries in Perthshire in 1981 (referred to as female 1, female 2 and male 1 respectively by Green *et al.* 1984); and (ii) J Twelves, tracking 7 wild otters on South Uist, Scottish Western Isles, again in 1981. Data were also obtained from a group of 3 otters released into a river on the Norfolk/Suffolk border on 5 July 1983 as part of a re-population trial (Jefferies *et al.* 1984b, 1986). These were females OT/1 and OT/2 and male OT/3 (Jefferies *et al.* 1985). The male was fitted with a radio-transmitter and tracked by myself, R M Jessop and A J Mitchell-Jones. Other notes are from

Plate 6. Male otter fitted with a harness and attached radio-transmitter to facilitate study of its behaviour in disturbance situations (Photograph Otter Trust)

surveys for spraints and holts or from long periods of observation and from the literature.

3.1.1 Bank anglers

Green *et al.* (1984), radio-tracking the male P3 on the River Earn in 1981, found that in rural areas anglers created the main potential for disturbance. This was especially so in July and August when all-night fishing for sea trout was at its peak, as this individual was almost completely nocturnal. Intensive fishing was restricted to just a few parts of the male's home range but all of these were situated within its centres of activity. A maximum of 12 night-anglers was recorded on one night on a 1500-m stretch of the river. Nonetheless the otter remained within this area, bypassing anglers either on land, through dense vegetation, or under water. Male P3 was tracked circumventing anglers on 21 nights. On a further 14 nights he was found to be active within 200 m of anglers. Although their presence often slowed his progress, only once were they seen to be responsible for turning him back. On one occasion, at sunset, this male was observed to swim unseen past one group of young children, 4 bankside anglers, one wading angler and a number of walkers, besides gliding beneath a busy bridge where the river narrowed and was only 15 cm deep.

On 3 August 1983, I began tracking male OT/3 in Suffolk after it emerged at 1920 hours GMT (dusk 1948 hours GMT) and started swimming purposefully upriver. This behaviour is quite different to that when feeding is the main object, when the otter traverses a particular reach several times over half an hour or more. After I had radio-tracked him for 1650 m, I decided to test his reaction to a person, such as an angler, standing still on the bank, by walking to a point ahead of him. I first observed his head as he swam along the far bank at about 50–60 m distance. Very shortly after, he obviously became aware of me although I was stationary. Instead of turning back, he crossed over to my bank and dived. I observed the coloured radio-box passing by beneath my feet at less than 1 m under water at 2016 hours GMT. The radio-signal continued as he swam upstream on the surface. Without the radio equipment, I would have been unaware of his presence on the river. The distance travelled that night was 7930 m, including a stretch under a busy main road bridge, suggesting that he had not been deterred at all from his original intention to explore upstream.

Macdonald and Mason (1983) have noted that some of the most expensive and carefully maintained game fisheries in Britain, eg on the Rivers Wye and Severn, are also the stretches of those rivers that are most extensively marked by otters. The same could be said for the Aberdeenshire River Dee (Jenkins & Burrows 1980). So, bank angling activity for game fish appears not to affect otter presence and marking behaviour adversely. On the other hand, although there was an even distribution of otter signs in the spring on a tributary of the upper River Severn, within 2 weeks of the opening of the coarse fishing season in mid-June otters ceased using, or passed through without marking, the central section of the river where the fishing rights were owned by an active angling club. An even distribution of spraints was resumed at the close of the fishing season in March (Macdonald *et al.* 1978). The difference may be due to the usually greater density of coarse fishermen compared to that of game fishermen. The last-named authors also observed that anglers favoured the cut-back trees and stumps as fishing sites, so probably deterring otters from using the root systems for holts.

3.1.2 Boat anglers

A young male otter with a radio-harness was released into Loch Leven (Fife) in early September 1985 by J and R Green of the Vincent Wildlife Trust. Loch Leven is intensively boat fished for trout, with 43 fishing boats fishing for 3 sessions a day. However, the otter avoided the boats by waiting for the opportune moment before moving. On one occasion, 2 boats were seen to pass within 100 m of an island on which the otter had a holt. He waited until they had both passed before immediately swimming for the shore, unseen by anybody on the boats (J Green, personal communication).

3.2 Reactions to angler-related disturbance

Angling is the most important recreational activity occurring along rivers, as measured by the number of people using that habitat. Sixty per cent of anglers travel to their fishing sites by car (Natural Environment Research Council 1971) and try to get their now heavy tackle as close to that site as possible. Macdonald *et al.* (1978) counted 159 cars parked by anglers along the River Teme on Sunday, 11 September 1977. These cars cause disturbance by noise and lights, too, if arriving early and leaving late. Associated camping and radios cause further noise and lights. Also, anglers walk the river banks when moving to their chosen site. The following collected examples, although not of disturbance due directly to angling, give an indication of the effect of this angling-related disturbance.

3.2.1 People walking along river banks

Nearly half of the known 29 resting sites used by male P3 in Perthshire were subject to some disturbance, usually by people walking by on the bank. Only on 3 occasions in the 98 days of tracking did the disturbance cause the otter to move to other sites. On the other hand, female P1 vacated an insecure holt within minutes of it being approached by people (Green *et al.* 1984).

On the evening of 20 July 1983 at 2000 hours GMT, I had been observing the holt used during the previous day by the radio-harnessed male OT/3 in Suffolk for 45 minutes. There had been no movement from or within the holt, although 2 combine-harvesters had moved down the farm track nearby. However, when I approached the holt myself and stood on it, he moved silently out and slipped down a dry ditch unobserved.

Weir (1984) noted changes in otter behaviour correlated with high levels of human activity on the Norfolk coast. During the peak holiday months of July, August and September, otters appeared to restrict their activities for much of the time to the quieter parts of their range and travelled less. Any sprainting sites in close proximity to areas with intense human activity were left unmarked at this time, although otters passed by them. The same places were used as marking points at other times of the year.

3.2.2 Noise and lights

Sharp and sudden noises cause instant flight to the nearest water, as was noted by Weir (1984) when an otter he was observing was disturbed by the noise of a falling bicycle. Once in the water, the behaviour is secretive for 20–30 min afterwards. An otter was recorded as swimming 140 m underwater on the River Stiffkey in Norfolk about 20 min after a firework display had ended upstream. Another otter gave a similar performance 30 min after 2 men had been shooting at night near to the river. A third left Cley Marsh when people were shouting 500 m away and swam the 110 m of the drain to the River Glaven underwater, after a boat had passed the drain entrance (Weir 1984).

Continuous noises, even loud machinery noises, are tolerated. Weir (1984) noted no disturbance from continuous loud fair music to 2 otters on Cley Marsh. Also, during July 1983, I have observed both female and male otters (OT/1, OT/2 and OT/3) passing by a continuously running tractor working an irrigation pump only 10 m from the river bank. One couch used by female P1 in Perthshire throughout one day was situated under a pile of rhododendron (*Rhododendron ponticum*) sticks on the verge of a busy main road (Green *et al.* 1984). Car headlights do not deter otters from crossing roads at night, and otters appeared oblivious to the village lights illuminating them on Cley Marsh (Weir 1984). On the other hand, lights accompanied by talking may evoke the usual reaction of swimming underwater to avoid detection. On the night of 23 July 1983 at 2145 hours GMT, male OT/3 was being radio-tracked in Suffolk as he approached a road bridge close to a village. Although this road usually bore little traffic, on this night a car with headlights on was parked there with people talking. I observed the otter dive as he approached the bridge and the faint 'V' surface wash as he swam under before surfacing on the downstream side and continuing his journey.

3.3 Reactions to other bankside disturbance

It would be unfair to suggest that anglers provide the only disturbance to otters on waterways. The walking of dogs, mink hunting, boats and bank maintenance by Water Authorities all contribute. The effects of these types of disturbance should be evaluated and compared to that produced by anglers. Also, the ability of the otter to cope voluntarily with urban situations gives a useful indication of its tolerance to disturbance.

3.3.1 Dogs

There is strong evidence that otters regard dogs as a higher potential danger than man. Although, as noted in 3.2.1 above, male P3 showed only occasional reactions to people walking near its resting sites, it reacted adversely on 2 of only 3 occasions when dogs were in close proximity. On one instance, the otter abandoned the holt for 3 weeks, using a new site 200 m distant (Green *et al.* 1984). Similarly, with J Twelves I have observed radio-harnessed otters on South Uist rolling on seaweed couches and sleeping at mid-day, apparently feeling secure within 100 m of people digging peat and taking very little notice of them. This behaviour changes immediately to alertness when a dog barks at a nearby house. Otters abandoned a favoured site for several months in Norfolk after a pack of otter hounds passed through (Macdonald *et al.* 1978).

3.3.2 Boats

There is little information on the reaction of otters to boats, except that contained in Sections 3.1.2 and 3.2.2.

3.3.3 Riverside working activity

Weir (1984) observed that construction work on the River Stiffkey sluices in Norfolk, involving heavy equipment, considerable habitat change and taking place over a year, caused a large reduction in marking behaviour at the site. Spraints were not found regularly at that site again until 2 months after completion. On the other hand, minor maintenance work, such as cutting bankside vegetation, had no effect on otter movements. Green *et al.* (1984) found that fencing operations within 30 m of a holt occupied by female P1 caused no detectable effect on the otter inside, not even causing any fluctuation in the radio-signal. Similarly, tree felling over several days near to a holt occupied by the same female only caused her to abandon it when the foresters were directly overhead. However, both these holts were deep and secure.

3.3.4 Otters in urban situations

Of 32 urban and rural resting sites used by male P3 and female P1 in Perthshire, 15 were situated within 250 m of human habitation, compared with 17 that were 251–850 m distant. Thus, there was no evidence that they deliberately chose sites distant from habitation. In fact, the most frequently used of all 47 known resting sites was a holt near a cafeteria with outdoor tables. Journeys through the 2 towns and one village within the range of male P3 occurred regularly, up to 4 times per week. A total of 39 urban journeys was recorded during 98 days of monitoring. Urban activity was observed from 30 min before sunset until 3 h after sunrise, so occasionally occurred during periods of human activity. In quieter sections of urban habitat, such as gardens, he travelled on the surface. Elsewhere, he travelled underwater, either in mid-current or beneath overhanging vegetation. At such times he was seldom seen, even by his trackers (Green *et al.* 1984). Macdonald and Mason (1983) examined 50 5-km stretches of rivers in Wales and the west Midlands. On these stretches they measured the effect of 2 features of potential disturbance: (i) the presence of camp/caravan sites on the bank, and (ii) human settlements, on marking by otters. Neither feature had a significant effect on the intensity of marking with spraints.

There are many examples of otters actually living in or passing through very large urban areas. In Aberdeen and Glasgow, Green and Green (1980) found spraints in the suburbs, as did Chapman and Chapman (1982) surveying in Dublin and Athlone. Macdonald and Mason (1980) found spraints within the town of Ullapool (Easter Ross), where otters scavenged from the fish market and from moored fishing vessels. Two regularly used holts were found well within town boundaries in Wales and west Midlands (Macdonald & Mason 1983). Two otters used to inhabit the Mill Pond in Pembroke town (Sansbury 1957) and 2 pairs had holts in the centre of York on the Rivers Ouse and Foss for many years (Taylor 1956).

3.4 Differences between the sexes and at breeding

3.4.1 Differences between the sexes

Most of the radio-tracking work (Green *et al.* 1984; Jefferies *et al.* 1986) has been carried out on males. It is impossible to be sure of the sex of an unmarked otter in the field, unless it is with cubs. Also, much of the site marking with spraints is carried out by the males (Green *et al.* 1984; Hillegaart *et al.* 1985) so well-marked sites in towns are not necessarily occupied by both sexes. Thus, one must be careful not to form an opinion on the effects or non-effects of disturbance on otters from information from one sex only. Green *et al.* (1984) noted that male P3 was markedly bolder than either female P1 or P2 and that he often used disturbed habitat. One day he moved voluntarily from undisturbed woodland to an urban couch bordering a village street in daylight after 0900 hours. On the other hand, female resting sites were invariably situated in the most secure habitat available; female couches were covered with dense scrub, whilst male couches were often exposed. Female otters used twice as many underground holts (68%) as surface couches (32%), whereas males showed the reverse trend (couches 67%; holts 33%). Also, females P1 and P2 tended to avoid areas of relatively high disturbance. These factors suggest that the female has a higher sensitivity to disturbance than the male.

3.4.2 Breeding females

The holts I have seen that have been chosen and used by the female to give birth and to rear her cubs are the most secure and/or secluded of all holts, often in rock clefts. Green *et al.* (1984) and Harper (1981) also refer to the security of breeding holts. Security is enhanced by the reduction or cessation of spraint marking around the area (Östman *et al.* 1985). All this suggests that the female is most sensitive to disturbance at this time, even that occurring some distance away from the site. This suggestion is supported by the following 2 examples. (i) Weir (1984) noted that, when Salthouse Marsh (Norfolk) was regularly shot over, a female otter with cubs had never been known to lie up there, always preferring the neighbouring Cley Marsh Nature Reserve. However, when shooting ceased in 1969–70, a female with cubs was lying up on the marsh more often than it was using Cley. Salthouse was then quieter than Cley. When shooting resumed, otter families were no longer recorded there. (ii) On 30 October 1981 at 1220 hours GMT, I was observing a radio-harnessed male otter near Loch Carnan, South Uist, with J Twelves and A Mitchell-Jones. This animal swam to where we were partially concealed and then came ashore only 7 m away, having caught a butterfish (*Pholis gunnellus*) which he proceeded to eat. At this point, we noticed that a second otter had followed him across the loch. The second otter, which we took to be a female from what followed, was much more wary and noticed us immediately. After making the usual aggressive 'huffs', it came ashore to make sure of us before diving and swimming back to the island from which it had come. The male, after coming within 2 m of us without alarm, then walked away. The female entered a secure rock holt, some 130 m away, and brought out a single, very young cub. Her concern that we were so near to the breeding holt caused her to move this cub out of sight around the headland. The cub could not swim more than 2 m before resting on the beach, so progress was slow.

On the other hand, it must be said that there are several examples of females producing and rearing cubs in disturbed situations. A litter of otter cubs was reared in a holt below one of the busy main jetties at the Sullom Voe Oil Terminal on Shetland (Berry 1985). Otters regularly raised a family within the York city boundary, using an old drain near Ouse Bridge (Taylor 1956). At Horsey Staithe in Norfolk, 2 otters with 3 cubs using the Mill Dyke were said to take no notice of the millman, even sitting on his boot (Banham 1966).

4 Various other interactions

4.1 Fish traps

Underwater fish traps and nets such as eel fykes set in fresh and brackish waters in Britain are known to have caused the drowning of at least 48 otters in the 9 years from 1975 to 1984 (Jefferies *et al.* 1984a). Most of this activity is commercial fishing for food, but some fykes and traps are set to catch fish because of angling interests, either to catch coarse fish for relocation or to remove predators.

4.2 Fish farms

The majority of game fish farming is done for the table, but some is to provide fish for release for sporting purposes. One problem of concentrating fish in a small volume and siting them near otter-inhabited rivers and coasts is that they form an obvious attraction for all aquatic predators. Problems arise when the owner wants to trap, remove or kill otters from the neighbourhood. Such a site could then become a form of 'sump and drain', continually attracting otters to an empty but good fishing area and then removing them.

4.3 Habitat modification

Some modification of the habitat occurs wherever anglers need to get to the water side, ranging from removal of the emergent vegetation and the formation of a bank 'notch' every 7 m or so, to the removal of trees to provide space for casting. Cover is essential to the otter and the roots of bankside trees provide holts.

4.4 Night angling

Otters are largely nocturnal on mainland Britain, particularly in the south. Those radio-tracked in Perthshire and Suffolk in the studies noted above usually emerged within an hour of dusk, with emergence times changing with dusk times (Green *et al.* 1984; Jefferies *et al.* 1986). This behaviour may be a long-term reaction to continuous daytime disturbance. However, Jenkins (1980) concluded from his observations at Dinnet Lochs (Aberdeenshire) that otters were mainly nocturnal when at low population densities and much more diurnal when numbers were high. This conclusion would certainly fit with the observation that otters may be seen most often in daylight in the Northern and Western Isles, where density is high. Normally, largely nocturnal activity would tend to reduce the contacts between people and otters. On the other hand, it suggests that those areas where night fishing for carp (*Cyprinus carpio*) or sea trout (*Salmo trutta*) is practised on the mainland may be the places where interactions with otters are most likely to occur.

4.5 Fishing tackle and direct interaction

Pike (1950) referred to an occasion when an angler, who was fishing for pike (*Esox lucius*) with a live bait, caught an otter that had become hooked after taking the bait. Also, Reynolds (1985) has noted finding a large fish hook in the stomach of an otter road casualty from Orkney.

5 Conclusions

The available information shows that, generally, the otter is indeed very shy in that it takes great care to remain unobserved by humans. However, it is known to use intelligent means of getting round forms of disturbance rather than retreating from them. The usual method is either to use dense cover or to swim underwater, depending on what is to be traversed. With a maximum diving time of 3½ min (Jefferies *et al.* 1984a), and a swimming speed of 1472 m/hour (Jefferies *et al.* 1986), an otter may be able to cover around 80 m underwater before surfacing again, which would suffice for avoiding most problems. The fact that the otter is largely nocturnal on the mainland (Green *et al.* 1984) helps in this concealment. Using concealment, the otter will pass through towns and by anglers. Continuous noises, even loud machinery noises, and lights are tolerated, but sharp and sudden noises cause instant flight and secretive behaviour within that area for at least 20–30 min afterwards. People are tolerated quite close to holts, particularly if the latter are felt to be secure. Dogs, on the other hand, are not tolerated and will cause immediate withdrawal. By the areas they frequent and the shelters they choose, the females appear to be less tolerant of disturbance than the males. Females with cubs are not tolerant of disturbance at all. Much of the marking by otters on busy waters may be by males.

The above suggests that disturbance may not be the problem to otters that was at first envisaged, at least where there is sufficient cover left for them to move about secretly. If disturbance increases and cover is decreased, it seems likely that, at present densities, the otter would retreat from the disturbed area and would feed elsewhere, which could effectively reduce the area available for inhabitation by otters, and so presumably control the population size. On the other hand, in certain situations, such as the Northern and Western Isles, where otters live a semi-marine existence, at much higher densities than on the mainland, and are active throughout the daylight hours, they appear to be able to get accustomed to high disturbance levels and live close to people. This adaptability may be a necessity and a result of high population density. However, there is insufficient information to show how representative this behaviour is and whether otters would behave similarly in the rivers of the south, if the population ever reached such densities.

Otherwise, the main problem from disturbance would appear to be the potential effect on breeding females. Sufficient quiet and secure places are required to be protected to enable the otter to rear cubs in seclusion. The Wildlife and Countryside Act 1981 makes it illegal to disturb such places intentionally, but cannot protect them from accidental disturbance. Fortunately, most of the places I have seen used by breeding otters are away from main rivers and so overgrown as to be unlikely to be used by anglers. Otters can and do breed in every month of the year (Stephens 1957), so none of the anglers' close seasons provides much benefit in terms of a reduction in potential disturbance.

Anglers are no different to any other group using the river banks when it comes to causing disturbance, except that they form the majority of such people and are present for much of the year, even in winter. On the other hand, the evidence suggests that standing people would, in fact, produce less disturbance than those walking or exploring, in that they are easy to observe and pass and they disturb a shorter distance of waterway. They provide much less disturbance than people walking their dogs.

With regard to the other interactions mentioned in Sections 2 and 4: (i) it is unlikely in these more enlightened times that the killing of otters (now illegal) to protect fisheries is going to be the important mortality factor in the future that it was in the past (although it may still occur); (ii) research is in progress at the moment to find methods of excluding otters from fish traps without impairing their efficiency; (iii) fish farms can deter otters by the use of guard dogs or, in certain places, electric fencing; (iv) the habitat modifications and cover removal carried out by Water Authorities to improve drainage far outweigh those made by anglers (Macdonald 1983); (v) although, when night fishing, humans have lost the use of sight and so are less observant, anglers are just as obvious to the otter as they would be in daylight and just as easily avoided; (vi) finally, I do not think that such direct interactions as mentioned in Section 4.5, or the consumption of fish bearing old hooks, have happened very often, as I have never found any evidence in the bodies of any of the 80 otters I have dissected.

On the positive side, anglers are always vigilant for pollution incidents and form a powerful lobby for cleaner rivers. Pollution is thought to have been the main cause of the otters' decline (Chanin & Jefferies 1978) and is still of concern in fresh waters (Jefferies 1985). After short-term pollution incidents, such as those causing deoxygenation, waters are restocked for angling, coincidentally ensuring a food supply for the otter. An adult otter consumes about 1 kg of fish a day (Stephens 1957) but, as it takes it from around 25 miles of waterway (Green *et al.* 1984; Jefferies *et al.* 1986), the effect is negligible. Otters and anglers can co-exist in peace.

6 Summary

Angling interests, and the consequent heavy pressure of predator control, were one of the main causes of the reduction of the otter (*Lutra lutra*) population during the 19th century. As with other carnivores, they appear to have regained their numbers during the first half of the 20th century with the reduction in 'keeper-ing. This improvement was followed by the sudden and catastrophic crash in the British otter population starting in 1957–58, which occurred simultaneously over most of the country but was most severe in the south and east. Analysis of the available information does not implicate angling, disturbance or fishing interests generally as the main causative agents of this recent and continuing decline. The otter is apparently a shy and secretive animal, difficult to watch and study. Knowing this, conservationists have been worried about the increasing disturbance caused by the growing use of waterways for pleasure boating and angling. However, examination of sighting and survey data suggests that these early fears may have been exaggerated and the tolerance level higher than originally thought, at least where there is sufficient cover left for otters to move about secretly. Also, the use of radio-transmitters and the tracking of male otters through actual disturbance situations have supported this conclusion by demonstrating how they react and how they cope with people and their machines, largely by swimming underwater. There is less information on females which appear to be more sensitive, particularly when with cubs. Thus, the main problem from disturbance would appear to be the potential effect on breeding females. Sufficient quiet and secure places must be protected to enable the otter to rear her cubs in seclusion.

7 References

Banham, P. R. 1966. Norfolk Mammal Report 1965. *Trans Norfolk Norwich Nat. Soc.*, **20**, 303–311.

Bell, T. 1837. *A history of British quadrupeds.*London: Van Voorst.

Berry, R. J. 1985. *The natural history of Orkney.* London: Collins.

Chanin, P. R. F. & Jefferies, D. J. 1978. The decline of the otter *Lutra lutra* L. in Britain: an analysis of hunting records and discussion of causes. *Biol. J. Linnean Soc. Lond.*, **10**, 305–328.

Chapman, P. J. & Chapman, L. L. 1982. *Otter survey of Ireland 1980–81.* London: Vincent Wildlife Trust.

Crawford, A. K., Jones, A. & McNulty, J. 1979. *Otter survey of Wales 1977–78.* Lincoln: Society for the Promotion of Nature Conservation.

Cuthbert, J. H. 1973. Some observations on scavenging of salmon *Salmo salar* carrion. *West. Nat.*, **2**, 72–74.

Green, J. & Green, R. 1980. *Otter survey of Scotland 1977–79.* London: Vincent Wildlife Trust.

Green, J., Green, R. & Jefferies, D. J. 1984. A radio-tracking survey of otters *Lutra lutra* on a Perthshire river system. *Lutra*, **27**, 85–145.

Harper, R. J. 1981. Sites of three otter (*Lutra lutra*) breeding holts in fresh water habitats. *J. Zool.*, **195**, 554–556.

Hillegaart, V., Östman, J. & Sandegren, F. 1985. Area utilisation and marking behaviour among two captive otter (*Lutra lutra* L.) pairs. *Otters, Otter Trust, 1984*, 64–74.

Howes, C. A. 1976. The decline of the otter in South Yorkshire and adjacent areas. *Naturalist*, **101**, 3–12.

Jefferies, D. J. 1985. An otter casualty from Breydon Water, Norfolk. *Otters, Otter Trust, 1984*, 23–24.

Jefferies, D. J., Green, J. & Green, R. 1984a. *Commercial fish and crustacean traps: a serious cause of otter* Lutra lutra *(L.) mortality in Britain and Europe.* London: Vincent Wildlife Trust.

Jefferies, D. J., Jessop, R. & Mitchell-Jones, A. J. 1984b. Re-introduction of captive-bred European otters *Lutra lutra*, to suitable field situations in England. *Otters, Otter Trust, 1983*, 37–40.

Jefferies, D. J., Wayre, P., Jessop, R. M., Mitchell-Jones, A. J. & Medd, R. 1985. The composition, age, size and pre-release treatment of the groups of otters *Lutra lutra* used in the first releases of captive-bred stock in England. *Otters, Otter Trust, 1984*, 11-16.

Jefferies, D. J., Wayre, P., Jessop, R. M. & Mitchell-Jones, A. J. 1986. Reinforcing the native otter *Lutra lutra* population in East Anglia: an analysis of the behaviour and range development of the first release group. *Mammal Rev.*, **16**, 65–79.

Jenkins, D. 1980. Ecology of otters in northern Scotland I. Otter (*Lutra lutra*) breeding and dispersion in mid-Deeside, Aberdeenshire in 1974–79. *J. Anim. Ecol.*, **49**, 713–735.

Jenkins, D. & Burrows, G. O. 1980. Ecology of otters in northern Scotland III. The use of faeces as indicators of otter (*Lutra lutra*) density and distribution. *J. Anim. Ecol.*, **49**, 755–774.

Langley, R. J. W. & Yalden, D. W. 1977. The decline of the rarer carnivores in Great Britain during the nineteenth century. *Mammal Rev.*, **7**, 95–116.

Lenton, E. J., Chanin, P. R. F. & Jefferies, D. J. 1980. *Otter survey of England 1977–79.* London: Nature Conservancy Council.

Macdonald, S. M. 1983. The status of the otter (*Lutra lutra*) in the British Isles. *Mammal Rev.*, **13**, 11–23.

Macdonald, S. M. & Mason, C. F. 1980. Observations on the marking behaviour of a coastal population of otters. *Acta theriol.*, **25**, 245–253.

Macdonald, S. M. & Mason, C. F. 1983. Some factors influencing the distribution of otters (*Lutra lutra*). *Mammal Rev.*, **13**, 1–10.

Macdonald, S. M., Mason, C. F. & Coghill, I. S. 1978. The otter and its conservation in the River Teme catchment. *J. appl. Ecol.*, **15**, 373–384.

Mitchell-Jones, A., Jefferies, D. J., Twelves, J., Green, J. & Green, R. 1984. A practical system of tracking otters *Lutra lutra* using radio-telemetry and 65-Zn. *Lutra*, **27**, 71–84.

Natural Environment Research Council. 1971. *National angling survey 1970.* London: NERC.

Nethersole-Thompson, D. 1951. *The greenshank.* London: Collins.

O'Connor, F. B., Sands, T. S., Barwick, D., Chanin, P., Frazer, J. F. D., Jefferies, D. J., Jenkins, D. & Neal, E. G. 1977. *Otters 1977.* London: Nature Conservancy Council/Society for the Promotion of Nature Conservation.

Östman, J., Hillegaart, V. & Sandegren, F. 1985. Behavioural changes in captive female otters (*Lutra lutra* L.) around parturition. *Otters, Otter Trust, 1984*, 58–63.

Pike, O. G. 1950. *Wild animals in Britain.* London: Macmillan.

Potts, G. R. 1980. The effects of modern agriculture, nest predation, and game management on the population ecology of partridges (*Perdix perdix* and *Alectoris rufa*). *Adv. ecol. Res.*, **11**, 1–79.

Reynolds, P. 1985. Fish hook in otter stomach. *Otters, Otter Trust, 1984*, 25.

Sansbury, D. G. 1957. Field notes – mammals. *Nature Wales*, **3**, 404–405.

Stephens, M. N. 1957. *The natural history of the otter.* London: Universities Federation for Animal Welfare.

Taylor, E. W. 1956. A summary of our knowledge of Yorkshire mammals 1881–1955. *Naturalist*, **81**, 37–44.

Weir, V. 1984. *The otter.* London: Vincent Wildlife Trust.

Habitat modification associated with freshwater angling

K J MURPHY* and H G PEARCE†
*Department of Botany, University of Glasgow, Glasgow
†Lee Valley Regional Park Authority, Bulls Cross, Enfield

1 Introduction

Freshwater angling is a major and economically important leisure activity in the British Isles. In 1980, it was estimated that there were 2.8 million freshwater anglers in England, Wales and Scotland: 0.7 million fished for game or salmonid species, and 2.1 million fished for coarse fish or non-salmonid species (National Opinion Polls 1980).

In recent years, growing concern has been expressed by various groups about the impact of anglers and angling activity on wetland habitats and their associated wildlife (Mills 1979; Edwards & Bell 1985, 1987). The game and coarse fish species of recreational importance are widely distributed in England, Wales, Ireland and Scotland (Maitland 1972). Freshwater angling occurs on the majority, if not all, of the linear and enclosed freshwater habitat types in Great Britain, including rivers, streams, canals, dykes, drains and ditches, lochs, tarns, lakes, ponds, pools, flooded gravel pits and reservoirs. With up to one million angler visits being made each week to these habitats during the open fishing seasons for coarse and game fish (National Opinion Polls 1980), and with various environmental modifications being undertaken by angling and fisheries management organizations, the concern about habitat modification is not surprising.

In contrast to other forms of outdoor recreation, the ecological effects of which have been the subject of considerable research (International Union for the Conservation of Nature and Natural Resources 1967, Satchell 1976; Salanki & Biro 1979; Tivy & O'Hare 1981), the impact of freshwater angling on the environment has been poorly studied.

In this paper, habitat modifications associated with freshwater angling activity and fisheries management practices on angling waters are examined.

2 Habitat modification due to angling activity

Habitat modification due to angling activity occurs because of the desire of anglers to gain access to and exploit a desirable fishery resource. Angling activities that can lead to habitat modification can be divided into 2 groups (Liddle & Scorgie 1980): those that are undertaken deliberately, and those that are inadvertent or accidental. The clearance of bankside vegetation along with the digging out of banks is frequently undertaken by anglers, particularly at coarse fisheries, to improve rod and line access and to create comfortable fishing positions to lay out their equipment. Liddle and Scorgie (1980) describe a section of the River Ouse near Huntingdon where, adjacent to an anglers' access track, some 30% of the bank vegetation had been modified from tall bank species to a short sward of perennial rye-grass (*Lolium perenne*), smooth meadow-grass (*Poa pratensis*), greater plantain (*Plantago major*) and knotgrass (*Polygonum aviculare*). At a distance of 300 m from the track, 20% of the banks were similarly affected. Plant species diversity was found to have increased in this study, but the fragmentation of the bankside vegetation continuum is not usually beneficial to river corridor wildlife. In rural areas, anglers who drive to and park adjacent to water areas can create rough roads across the countryside and flatten bankside vegetation. The impact of vehicles on the terrestrial environment has been described in detail by Tivy and O'Hare (1981), and thus is not considered here. The impact of anglers' vehicles is similar to that of other vehicle operators on the same terrain. Dense footpath networks may be created by anglers walking along or around the banks of a waterbody. In creating such access routes, vegetation is trampled and bushes, shrubs and trees may be physically damaged, with low branches being broken off.

Rees and Tivy (1978) describe the impact of anglers and other recreational users on Scottish lochshore vegetation in producing 'Indian file' tracks some 0.2–0.5 m wide through reedswamps. Typically, such footpaths ran parallel to the shore at the junction of 2 different plant communities; vegetation on either side of the paths was little affected (Figure 1). On little-used paths, reed (*Phragmites australis*), reed canary-grass (*Phalaris arundinacea*), reed sweet-grass (*Glyceria maxima*) and sedge (*Carex* spp.) were only slightly damaged, whilst on medium-use paths these species were replaced by harder-wearing species including bent (*Agrostis* spp.) and meadow-grass (*Poa* spp.), amphibious bistort (*Polygonum amphibium*), knotgrass or forget-me-not (*Myosotis* spp.). The heavily used paths largely consisted of bare substratum and occasional invading species. Rees and Tivy (1978) also characterize the relative trampling susceptibility of different lochshore species. In general, the taller grasses and other emergents growing on wet substrates, eg *Phragmites*, were more susceptible than shorter, tougher plants found higher up the shore, eg *Carex* spp.

Sukopp (1971), who studied the impact of intense recreational use of the shores of the Berlin Havel River system, noted that on this particular system anglers had only a relatively small impact on the shoreline. Other recreational users, particularly boaters, bathers and swimmers at densities of up to 9 people per metre of bank, had caused the loss of one-third of the fringing

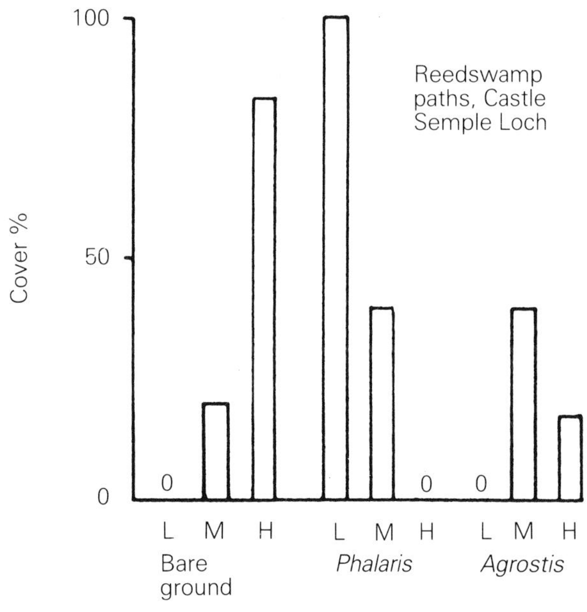

*Figure 1. Effects of trampling at low, medium and high densities of 'visitor passages', recorded by pressure-sensitive recorders, on cover of reed canary-grass (*Phalaris arundinacea*), creeping bent (*Agrostis stolonifera*), and bare ground, within reedswamp paths*

Intensity of trampling (passages/week):

L = low:	*2–3*
M = medium	*c13*
H = high	*c30*

(source: Rees & Tivy 1978)

reed community, over a 5-year period (1962–67). Under this intensive recreational use, a species-rich hydrosere community of submerged/floating/reedswamp/willow carr vegetation was replaced by a low cover of vegetation or by bare ground.

Bankside vegetation can also be affected by 'brolly tents' and anglers' camping activities, though the extent of vegetation flattening caused by this activity and its effects have not been studied in detail.

Fire lighting by anglers fishing at night or during camping trips can result in damage to trees by anglers searching for firewood, and may result in bare patches of burnt unvegetated soil close to the water's edge.

By standing or sitting close to the water's edge, anglers frequently damage the marginal plant community and with intensive use can cause serious bank erosion (Plate 7i). The Dartington Amenity Research Trust (1973) found that 26% of the total perimeter of one Rickmansworth gravel pit used for coarse angling had been worn bare by angling usage at a rate of some 33.33 fishermen per 100 m of bank (per week) in summer. Anglers who stand or wade in shallow water can also disturb marginal sediments or gravel deposits and other bed materials, and occasionally uproot emergent, floating-leaved and submerged aquatic plants. The disturbance of soft sediment can lead to an increase in turbidity. Upon the resettlement of sediment, aquatic plants and invertebrates may get

Plate 7. Erosion effects on bankside habitat, before and after reinstatement:
i. eroded river bank used by coarse anglers

smothered in fine particulate matter which may have a detrimental effect on photosynthesis and feeding respectively.

Where boats are used for fishing, their launching, movements, anchoring, mooring and landing can erode banks and damage aquatic and marginal vegetation. Descriptions of such effects can be found in Sukopp (1971) and Rees and Tivy (1978), and are summarized by Liddle and Scorgie (1980).

Occasionally, anglers' hooks and other tackle get entangled in aquatic vegetation. It is likely that the extent of plant uprooting caused by anglers extricating tackle is relatively small in most situations, and more usually the angler loses the entangled tackle. The habitat effects of angler groundbaiting, deposition of litter and tackle loss can be considerable, and are discussed elsewhere in these proceedings. Another form of habitat modification from angling activity can occur when toilet facilities are lacking; angler fouling (urinating and defaecating) can lead to amenity problems and artificial eutrophication, particularly of remote oligotrophic waters. For this reason, angling is banned at many reservoirs because of possible public health risks resulting from contamination of the supply of drinking water.

From the preceding paragraphs, it is evident that angling activity probably causes more detrimental than beneficial effects. The principal environmental component affected by angler activity is vegetation, and in particular the reedswamp plant community, which is especially valuable for the provision of shelter, food and breeding habitats for other wildlife, including birds, mammals, amphibians and invertebrates.

The degree of habitat modification due to angling depends on the type of waterbody and the abundance of vegetation in and around that waterbody. Generally small, narrow or shallow, vegetation-rich natural waterbodies are more prone to habitat modification due to angling than large, wide or deep, vegetation-poor artificial waterbodies. Figure 2 provides a summary of the impacts of angling activity on different habitat types. Natural lowland waterbodies with soft margins and abundant reedswamp, subject to particularly high intensities of coarse angling, may be particularly susceptible to damage, compared with an upland salmon river, with little emergent vegetation and stony substrates used by relatively few, more mobile, game anglers. In the more vulnerable habitats used by coarse anglers, the spring close season is probably of considerable importance in minimizing the impact of anglers at a phase of the plant growth season when vegetation is particularly vulnerable to damage. The resulting benefit to nesting waterfowl and spawning fish probably reinforces arguments in favour of the continuation of the coarse fishing close season. Where rainbow trout (*Salmo gairdneri*) fishing is permitted during the close season, this protection is removed and increased modification of bankside vegetation communities probably results.

In considering the habitat modification caused by angling activity, it is important to note that one environmentally insensitive angler may cause greater modification than numerous environmentally conscious anglers, and, as demonstrated by Sukopp (1971), other water-based recreational activities are also a potential cause of habitat modification.

3 Habitat modification due to angling management and fishery management practices

Habitat modification due to angling management and fishery management practices results from the desire of riparian owners or angling organizations to improve or restrict access, or to manipulate the fishery so as to improve fish production and anglers' catches. As few waters have ideal access or are consistently naturally productive, many angling venues are so managed.

3.1 Management practices to improve or restrict angler access

At remote fisheries sites (particularly where poaching is a problem), measures to restrict access and increase security are often introduced, and may also improve the perceived angling quality of the site and its economic value. To install fences and gates, vegetation may have to be cleared and this can have a detrimental effect on associated wildlife. In contrast, the introduction of brambles (*Rubus fruticosus*), hawthorn (*Cra-*

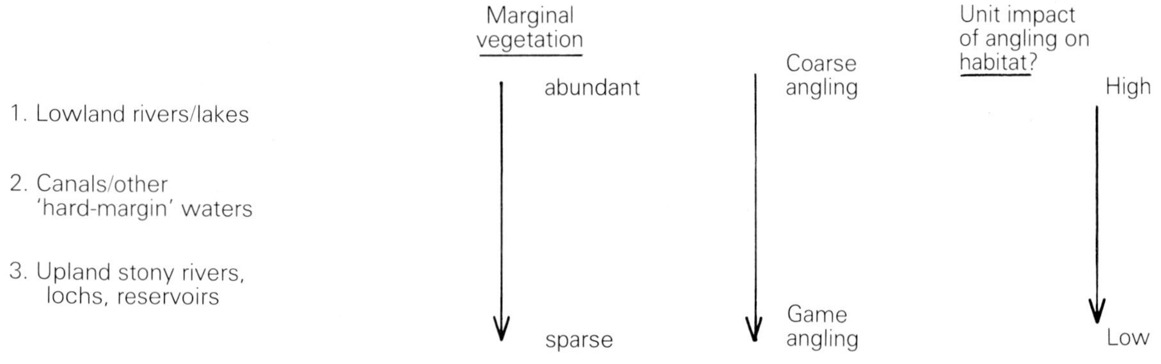

Figure 2. Hypothesized trends in habitat impact associated with different types of angling and vegetation, in relation to type of freshwater habitat

taegus monogyna) or fast-growing trees and shrubs as 'boundaries' can improve the wildlife value of a site through the provision of additional habitat.

A more frequent practice is to adopt measures to improve access for anglers. These measures may include the construction of roads, vehicle parking areas, footpaths, footbridges, boardwalks, steps down banks, fishing stances, platforms and cribs, boat launching slopes, jetties, moorings and landing sites. All these measures usually involve the clearance of vegetation, the addition of wood, stone, metal or other materials, and result in the modification of natural habitats, particularly during the period of installation. The construction of certain structures, such as footpaths, boardwalks and fishing stances (Plate 7ii), actually reduces habitat modification by anglers in the long term, by reducing the need for angling working parties for close season maintenance of banks and fishing positions. If undertaken regularly, these measures can cause habitat modification similar to that described earlier for angling activity.

3.2 Management practices to improve fishery production and anglers' catches

Management practices to improve fishery production and anglers' catches involve either physical, chemical or biological manipulation of the environment and biota of the fishery.

3.2.1 Physical manipulation

Physical manipulation of the environment can be used to create new fisheries (eg ponds, pools and gravel pits) and to modify existing habitats.

An example of the creation of a new fishery which not only produced a recreational amenity, but also increased the ornithological value of a low-grade pasture, is described by Bell *et al.* (1984). In total, 98 different species of bird (83 breeding species) were recorded around a newly created lake in south Wales. This type of management produces a rich and varied habitat attractive to other wildlife as well as fish. The trees planted by the angling club which managed the site were of particular value to wildlife.

Rivers are frequently subjected to maintenance activities (eg dredging) which can detrimentally affect fish habitat by physically altering the stream bed and channel sides. There is now a large literature on various mitigation techniques involving the use of structures that alter flow and water level, such as current deflectors and low weirs, which make up for the loss of natural features. For example, Swales and O'Hara (1983) describe how the use of low dams, current deflectors and artificial cover structures led to increases in fish population densities and biomass of 73% and 37% respectively, compared with pre-improvement values.

Plate 7. Erosion effects on bankside habitat, before and after reinstatement:
ii. reinstated bank with formed stance and regrown marginal vegetation along water's edge
(Photographs K J Murphy)

In the United States, Hunt (1968) recorded equally dramatic beneficial effects from the addition of bank cover and current deflectors to a trout stream. Brook charr (*Salvelinus fontinalis*) numbers, production, standing crop and yield increased by 156%, 17%, 78% and 196% respectively, following the installation of habitat improvement devices.

On enclosed waters, different types of habitat improvement devices are used. Fish-attracting and sheltering structures such as tyre or brushwood reefs are frequently introduced to large deep waterbodies such as reservoirs. These structures have the effect of concentrating fish and increasing the fish-holding capacity of these otherwise barren waterbodies (Parker *et al.* 1974).

3.2.2 Chemical manipulation

Many types of procedure are used, in different types of fishery waters, to modify their chemical characteristics. These measures fall into 2 broad categories – those aimed at restoring a degraded aquatic ecosystem to a point where a viable fishery is supportable, and those designed to improve or alter chemical conditions in an existing fishery, eg to maximize fish production.

3.2.2.1 Procedures of habitat restoration

Two examples of techniques falling under this heading are discussed here. In recent years, interest has increased in the problem of acidified waters (pH 4.0–5.0) located typically on hard, acidic rock catchments (eg granite) in upland areas, receiving highly acid precipitation. These waters are characteristically fishless, or support only low populations of small trout (Brown 1982; Harriman & Morrison 1982). One amelioration technique is to increase Ca^{2+} concentration and pH, by the addition of lime or other calcareous materials to the system or its catchment (Hultberg & Andersson 1982; Lindmark 1984). Such treatments have been applied to the catchments of acid lochs, such as Loch Dee in Galloway, Scotland (Burns *et al.* 1984) in recent years, with the intention of improving habitat conditions for salmonid breeding and survival, and have had broadly encouraging results.

In stratified hypertrophic lakes, or other systems in which dissolved oxygen availability is low, aeration techniques are an important habitat restoration procedure. It may be on an emergency basis (as in streams suffering sudden heavy organic pollution from silage effluent, many cases of which occurred in the area of the Clyde River Purification Board, in southern Scotland, during 1985). Alternatively, aeration may be part of a planned 'habitat rescue' operation, using large *in situ* pumps, to turn over or re-aerate waters of low oxygen status (Irwin *et al.* 1966; Lorenzen 1977a,b). Recent work has suggested that aeration may also have modifying effects on aquatic plant growth. Cooley *et al.* (1980) suggested that the growth of the submerged macrophyte *Hydrilla verticillata* was reduced by approximately 20% by an experimental aeration treatment, whilst nutrient regeneration from aquatic macrophytes into the water column may also be influenced by aeration (Ogwada *et al.* 1984). Laing (1979) claimed that, in combination with the usage of a phosphate-precipitation agent, aeration could significantly reduce submerged growth in treated waters. Aeration has furthermore been reported as increasing the efficacy of diuron treatments against filamentous algae in fish hatchery ponds (Kirby & Shell 1976). The precise mechanisms of these effects in aquatic plant habitats are poorly, if at all, understood, but aeration does show evidence of acting as a general synergist of aquatic weed control measures.

3.2.2.2 Habitat modification procedures influencing inorganic nutrient status of fishery waters

Fertilization, involving the addition of organic material (such as chicken manure) or inorganic fertilizer, is widely used to boost the productivity of fish, such as carp (*Cyprinus carpio*), in ponds and other standing waters in Europe (Wolny 1967). Nutrient-rich systems which may naturally support approximately 280–560 kg/ha/year of carp in ponds (Schaeperclaus 1933) may be boosted, by such additions, to 600–800 kg/ha/year carp production (Bregazzi *et al.* 1984). The increase in trophic status implied by such alterations to the chemical regime of treated ponds tends to be associated with quite marked alterations to both the algal and macrophyte communities of the waterbody (Vollenweider 1968; Newbold & Palmer 1979), and to the animal populations supported by the plant community. Although fertilization is not at present widely used to boost carp production in British waters, recent work (O'Grady & Spillett 1981; Bregazzi *et al.* 1984) suggests that there is a large untapped potential in southern British waters for such habitat alteration.

In contrast to fertilization procedures, nutrient stripping or nutrient immobilization techniques have been widely advocated on a more general basis to reduce the perceived problems of cultural eutrophication in fishery waters. An increase in plant nutrient availability associated with eutrophication is often blamed for increasing weed problems in waterbodies (Casey & Westlake 1974; Thomas 1978). It has consequently been suggested that limitation of major nutrient (N,P,K) concentration within the waters/sediment of a eutrophic system might limit plant growth (Dobolyi & Herodek 1980). The use of nutrient control agents (eg phosphate-precipitation agent $CaSO_4$:Al_2:$(SO_4)_3$:H_3BO_3) has been propounded (Martin *et al.* 1971; Laing 1979) for this purpose, as have various other techniques for nutrient stripping (using tertiary water treatment procedures) from point source inputs such as sewage outfalls. As techniques that may improve the site for wildlife as well as the fishery status of the treated system, such procedures are probably largely beneficial in their effects.

3.3 Biological manipulation

Discussion of the forms of biological manipulation

which may be applied to freshwater fishery habitats is limited here to the vegetation components of the ecosystem. Other aspects, such as predator control, are covered elsewhere in these proceedings. Vegetation management is a widely practised form of habitat management used in fishery waters, and may involve (i) reducing or increasing the overall quantity of vegetation present; or (ii) altering the structure of the aquatic plant community.

The aims of such habitat modification are typically one or more of the following.

i. To improve an angler's physical access to the water (for himself, his boat, or his line).
ii. To increase the level of habitat support offered to fish populations.
iii. To reduce the mortality risk for fish populations, eg providing shelter from predators in dense weed beds (Savino & Stein 1982), preventing low night-time dissolved oxygen levels, associated with the high net respiratory oxygen demand of dense weed growths during summer conditions (Brooker *et al.* 1977).
iv. For reasons of amenity (improving the aesthetic appearance of a fishery), or tradition ('we've always cut the weed').

Although excessive 'weed' growth is generally considered undesirable by anglers (Murphy & Eaton 1981), a moderate degree of plant cover is nearly always considered a valuable asset, in almost all types of freshwater fisheries (Bouquet 1978). Indeed, in certain fly fisheries, plant growth is considered indispensable (Wright 1973), because it harbours the nymphs and larvae that make the fishing technique possible. Higher plants, especially water-crowfoot (*Ranunculus* spp.), are particularly important in this context in chalk and limestone streams, lowland pasture streams, trout ponds, and small lakes or lochs. The epiphytic micro-organisms present on the leaves and stems of aquatic plants provide food for the invertebrates, which are in turn used as a major food resource by salmonids in such waters. In faster, rocky streams (eg upland and trout rivers), nymphs of insects (eg Ephemeroptera; Plecoptera) are primarily associated with the mosses and algae attached to the stones.

Coarse fisheries, too, undoubtedly benefit from the cover and increased invertebrate food supply provided by macrophytes, as well as from the provision of spawning sites in weed beds, for species such as roach (*Rutilus rutilus*) (Marshall & Westlake 1978; Keast 1984). Northcott (1979) showed the importance of aquatic plants in providing crustacean zooplankton as food for young roach. In static or slow-flowing waters, vegetation may particularly influence the overall structure of the invertebrate community (Dvorak & Best 1982; Scheffer *et al.* 1984). The attitude to vegetation management of chalk stream trout fisheries, paraphrased below from Wright (1973), summarizes an optimum approach to aquatic plants in freshwater fisheries management: '. . . the good manager looks on weed beds as essential regulators of currents, giving both fast- and slower-flowing areas. He plants and weeds as appropriate . . . controls the amount of bankside vegetation to give shelter to fish and camouflage to the angler, and to minimise erosion. He also recognises the aesthetic value of aquatic plants in their own right as part of the river ecosystem'.

3.3.1 Planting of aquatic macrophytes

In fishery waters deficient in macrophyte cover, plant management may involve the direct encouragement of macrophyte growth. The techniques are essentially long-established, but have re-emerged in Britain in recent years for economic as well as environmental reasons. Emergent macrophytes planted to control erosion (Bonham 1980), for general amenity reasons (Kelcey 1978), or directly for fisheries management purposes (Shireman *et al.* 1983), are all likely to improve the level of support for fish and other wildlife within the waterbody. Willen Lake, a balancing lake constructed in Milton Keynes, England, during the mid-1970s, is a typical example of an aquatic habitat used for a wide range of water-based recreation, including coarse angling, in which both submerged and emergent macrophyte plantings have been undertaken. Kelcey (1981) described some of the problems encountered in relation to the planting programme, which included the loss, by theft, of the attractive white water-lilies (*Nymphaea alba*), and the outcompetition of the planted submerged species by an aggressive invasive growth of Canadian pondweed (*Elodea canadensis*) in 1978. The resulting weed problem forced the managing authority to attempt submerged weed control measures, using the herbicide dichlobenil, in 1979–80. This action, in turn, produced a situation, undesirable in amenity, wildlife conservation and fishery support terms, of dominance by blue-green and filamentous green algae of the lake in 1981. Planted emergent macrophytes were established more successfully, but this case history shows some of the difficulties of managing aquatic vegetation to increase the diversity and abundance of the macrophyte community.

In the canal system of Great Britain, now used principally as a recreational and leisure resource (British Waterways Board 1980), the role of aquatic and bankside vegetation in improving both the level of amenity and of habitat support for wildlife is well recognized (Hall 1987). Angling is a major leisure use of the canal system, alongside pleasure-cruising (Inland Waterways Amenity Advisory Council 1975; Stabler & Ash 1977). Both anglers and conservationists appear to agree reasonably well with the British Waterways Board's (1981) practice of maintaining a moderate growth of aquatic vegetation in, and along the banks of, most canals, providing support for fish populations and wildlife, and improving the visual appearance of the watercourse (Murphy & Eaton 1981; Murphy *et al.*

1982). Powered boat traffic is, however, a major source of damage to macrophyte growth in heavily used canals, as well as a cause of bank erosion (Inland Waterways Amenity Advisory Council 1983; Murphy & Eaton 1983). A belt of emergent vegetation has long been recognized as an effective protection against bank erosion (Bonham 1980; Murphy *et al.* 1980), as well as providing some vegetation cover for other organisms in waterways otherwise deficient in macrophytes. The difficulty, however, lies in successfully establishing reeds, sedges or other emergents in a habitat which is continually disrupted by wave-scour caused by boat-wash. One solution, explored by the British Waterways Board in recent years (British Waterways Board 1982), has involved planting the lesser pond-sedge (*Carex acutiformis*) along the margins of certain canals, with reinforcement provided during the establishment phase by geotextile material. Clumps of sedge are planted in pockets of ballasted mesh fabric laid along the bank (Plate 8). The plants root through the fabric, which ultimately disappears from sight beneath the vegetation.

Plate 8. Plantings of greater pond-sedge (*Carex riparia*) into woven geotextile 'pocket' fabric for canal bank protection and habitat improvement purposes (Photograph British Waterways Board)

In comparison with the alternative measures (eg steel-piling) for bank protection in canals heavily used by boat traffic, such techniques offer a major improvement in wildlife support terms over the environmental sterility of a hard-edged waterway.

3.3.2 Aquatic weed control

Measures taken to reduce the growth of macrophytic vegetation in fishery waters fall under 3 main headings:

i. physical (manual or mechanical clearance);
ii. chemical (aquatic herbicides); and
iii. biological (eg introduction of plant pathogens or grazing organisms to control weeds).

Whatever type of weed control is used by the fishery manager, the ecological effects of treatment fall into the following groups:

i. effects on target weed plants (using the word 'weed' here in its true sense, as a growth of plants causing nuisance to people);
ii. direct effects on non-target organisms (eg toxicological side-effects of herbicide treatments);
iii. indirect effects on the aquatic ecosystem arising from the death of the target plants.

Clearly, this widely used group of habitat modification procedures may have important effects on both the fish and other wildlife populations of angling waters.

The overall effects of aquatic weed control on the target plant community of a waterbody are summarized in Figure 3 (from the excellent RSPB/RSNC Rivers and Wildlife Handbook: Lewis & Williams 1984). Weed control has the overall effect of returning the plant community to an earlier successional stage, but the community changes occurring during this process are often quite complex. Given time, the community normally recovers to its original hydroseral stage, irrespective of the type of weed control measure used (Wade & Edwards 1980; Wade 1982). However, repeatedly treated plant communities may be maintained for long periods in a species-poor, hydroserally early state, often dominated by filamentous algae and phytoplank-

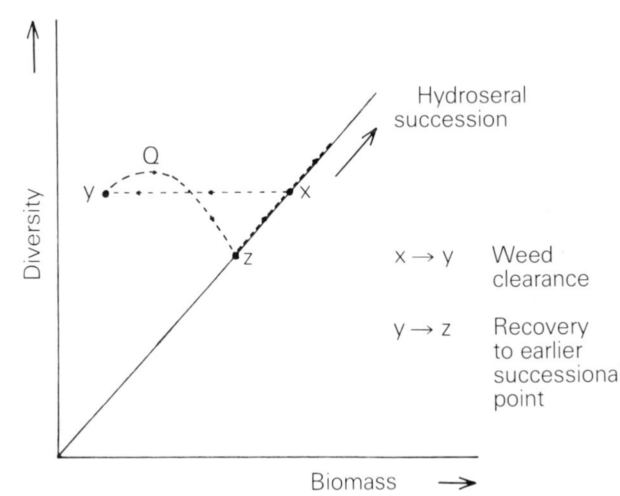

Figure 3. Hypothesized effects of weed clearance on the relationship between macrophyte community diversity, biomass and hydroseral succession. [$x \rightarrow y$: weed clearance phase; $y \rightarrow z$: recovery phase, to earlier successional stage; Q: point of maximal diversity reached during early phase of recovery of vegetation.] See text for further details (source: Lewis & Williams 1984)

ton. Such habitats as the Lincolnshire drainage ditches annually treated with dichlobenil, described by Harbott and Rey (1981), may be dominated by *Cladophora glomerata* or other algae, and support poorer invertebrate communities than ditches treated less frequently, using less efficient weed clearance procedures.

3.3.3 Physical weed control

Weed cutting is probably the oldest, and still the most widely used, of all aquatic weed control measures. It is typically carried out manually, using scythes or other tools, or mechanically, eg by specialized weed-cutting boats (Price 1981). Manual methods are slow, labour-intensive, and expensive, but are well suited to small-scale operations. It is easy to leave 'bars' of weed uncut (a common practice in chalk stream management (Birch 1976)), or otherwise manipulate plant growth as required (Soulsby 1974). Because not all the vegetation present is removed, regrowth is rapid. Dawson (1976a) has shown that in water-crowfoot streams, cutting stimulated the total annual vegetative production of the plant community, rather than causing any net reduction.

Weed cutting, of course, produces a degree of habitat disturbance, influencing non-target organisms as well as the target plants. Rabe and Gibson (1984) showed that the removal of submerged plants from a shallow lake caused a loss of substrate, food and shelter from predators, the net result being major changes in the distribution of invertebrate species.

In Danish streams, manual cutting of weeds was noted to cause a sharp increase in downstream drift of the shrimp *Gammarus pulex* (Kern-Hansen 1978), and may also directly influence fry mortality (Mortensen 1977). Swales (1982), however, suggested that it is difficult to separate the direct effects of weed clearance operations on fish populations from natural fluctuations in population density. Removal of weed to which spawn is attached, during spring in particular, is obviously potentially harmful to fish population recruitment. The effect of weed removal can also be indirectly damaging to non-target organisms. For example, weed removal from a river may result in a lowered water level, exposing spawn laid in the marginal shallows to the air, and increasing egg mortality rates for breeding populations (Mills 1981).

Mechanical cutting usually produces greater environmental disturbance than manual cutting. The scale both of this short-term disturbance and of longer-term habitat modification depends on the intensity and timing of the clearance operation. De Lange and van Zon (1978) suggested that relatively mild environmental impacts were associated with weed cutting (and also with the herbicide diquat). In Dutch drainage ditches, clearance during the summer months tended to improve the 'biological quality' (objectively assessed in terms of vegetational structure) of the aquatic habitat. An appropriate weed management regime of this type can effectively maintain the plant community at point 'Q' in Figure 3, ie a high diversity/low biomass plant community. Other forms of mild ecological 'pressure', such as a low density of boat traffic in a canal (Murphy & Eaton 1983), may have a similar effect on the vegetation, by maximizing the number of niches for plant species available within the aquatic ecosystem.

Dredging or other physical weed control measures that remove sediment and below-ground plant organs as well as plant foliage tend to be associated with a greater degree of habitat modification than straightforward cutting (Pearson & Jones 1975). Benthic invertebrate communities as well as open-water and epiphytic organisms are directly affected. George (1976) observed that the diversity/abundance of macro-invertebrates in dredged ditches was very closely related to stages of recovery of the plant community post-dredging.

Fish may also be trapped in the mass of plants removed by weed dredging operations. A study carried out in the USA (Haller *et al.* 1980) suggested that the 'replacement value' of fish destroyed by weed removal operations in a 65-ha lake during 1977 was as much as US$410,000.

Overall, although physical weed clearance in angling waters has a number of fairly well-documented short-term effects on habitat modification, the impact on the aquatic ecosystem is probably small in the medium to long term. A well-designed and correctly implemented weed control regime carried out for fisheries management purposes probably has a fairly mild impact on the wildlife and conservation value of a waterbody, and may be positively beneficial under certain circumstances.

3.3.4 Chemical weed control

Up to 1985, only a small number of herbicides had been approved for use in or 'near' fresh waters in Britain (Ministry of Agriculture, Fisheries & Food 1979) under the Pesticide Safety Precautions Scheme (PSPS) (Tooby 1978). The principal formulations used on submerged weeds are diquat (diquat-dibromide, a liquid formulation, and diquat-alginate, a gel formulation), dichlobenil and terbutryne (granular formulation). Emergent or marginal weed control mainly involves the use of dalapon, 2-4-D and glyphosate, all applied as liquid sprays (Plate 9).

In comparison with physical weed control measures, herbicides probably show, overall, a somewhat greater degree of direct impact on the aquatic habitat. The differences are:

i. herbicide treatment involves the deliberate introduction to the aquatic ecosystem of a toxin, of varying degree of toxicity to target and non-target organisms;
ii. plants are killed and decay *in situ*, leading to an

Plate 9. Selective control of emergent vegetation (branched bur-reed (*Sparganium erectum*)) along one bank of the Huddersfield Narrow Canal, using glyphosate, to improve water flow and access for anglers (Photograph K J Murphy)

increased organic loading on the system, with attendant water-quality problems;

iii. it is more difficult to limit the effects of treatment to the desired location (especially in flowing waters) because of the mobility of herbicide residues (Brooker & Edwards 1975).

Other effects, such as the loss of habitat for epiphytic organisms, are common to both physical and chemical control measures.

The effects of herbicides in aquatic habitats have been studied fairly intensively in recent years, because of the environmental dangers associated with chemical weed control. Two examples typify the potential problems involved, for one group of herbicides, the triazines, which act principally through photosynthetic inhibition in the target plants (Murphy 1982). In an experimental microecosystem containing submerged macrophytes, Wingfield and Bebb (1982) observed that terbutryne at 0.1 mg/litre active ingredient (the accepted field treatment rate) disrupted the normal diel cycle of dissolved oxygen and led to deoxygenation within a few days of treatment (Figure 4). The herbicide prevents photosynthetic oxygen release, but respiration of the plants is not initially interrupted, resulting in an imbalance of oxygen supply and demand within the system. As the herbicide breaks down or dissipates, so the oxygen regime is subsequently restored to near normal by the development of algal blooms, which use the nutrients released by macrophyte decay. The respiratory demand of decomposer organisms involved in the decay process exacerbates the oxygen sag, to an extent dependent upon the initial biomass of plants treated (Carpenter & Greenlee 1981). Similar effects with other herbicides, under field conditions, have frequently been noted (eg Brooker 1974).

Besides such indirect impacts of herbicide treatment on the treated ecosystem, there may be direct toxicity impacts on non-target ecosystem components. Under the PSPS clearance system, considerable emphasis is laid on toxicity testing, and only those herbicides with an acceptable margin of safety between LC_{50} and normal field use concentration are cleared. Figure 5 shows toxicity data for a triazine herbicide, cyanatryn (no longer used for aquatic weed control in Britain), in relation to effects on swimming activity in *Daphnia pulex* (Cladocera). The recommended dose rate of cyanatryn under field conditions was 0.1–0.2 mg/litre, maintained for up to 14 days by the slow-release formulation of the herbicide. The safety of the herbicide to *Daphnia* appears, from these data, to be at best only marginal.

In general, however, herbicides passed under PSPS for use in British fresh waters are (with the exception of dichlobenil which has a relatively high acute toxicity to fish and also causes sublethal damage to gills (Tooby & Spencer-Jones 1978)) probably adequately safe for use in fisheries in terms of direct toxicity. There are few, if any, documented examples of fish or other wildlife deaths caused by direct poisoning by PSPS-cleared, and correctly used, aquatic herbicides in Britain.

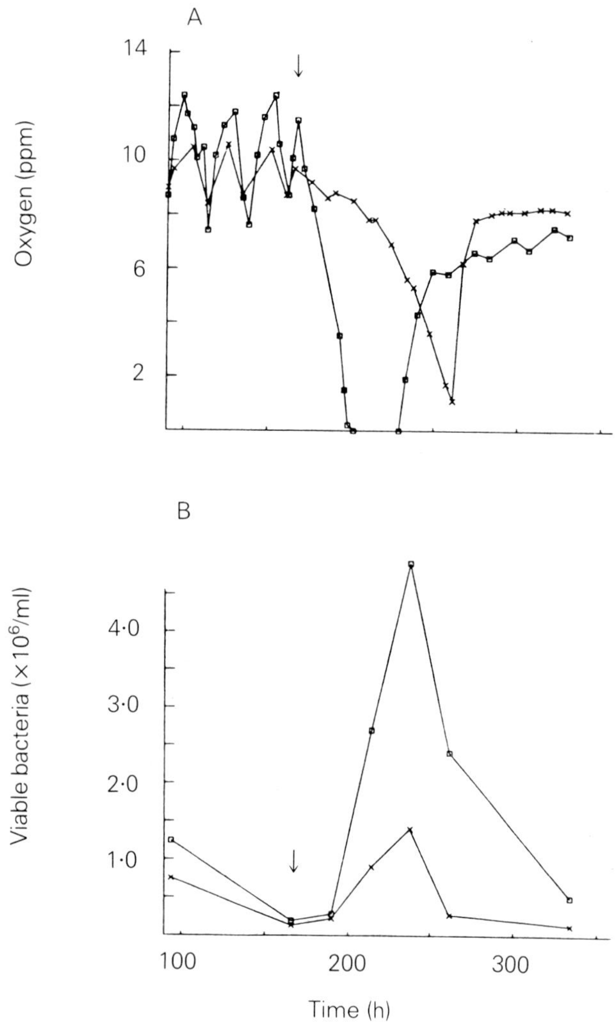

*Figure 4. Effects of terbutryne treatment (arrowed at 169 h after treatment) of Canadian pondweed (*Elodea canadensis*) (x) and rigid hornwort (*Ceratophyllum demersum*) (□) on: A. oxygen concentration, B. counts of viable bacteria, in an experimental microecosystem. (source: Wingfield & Bebb 1982)*

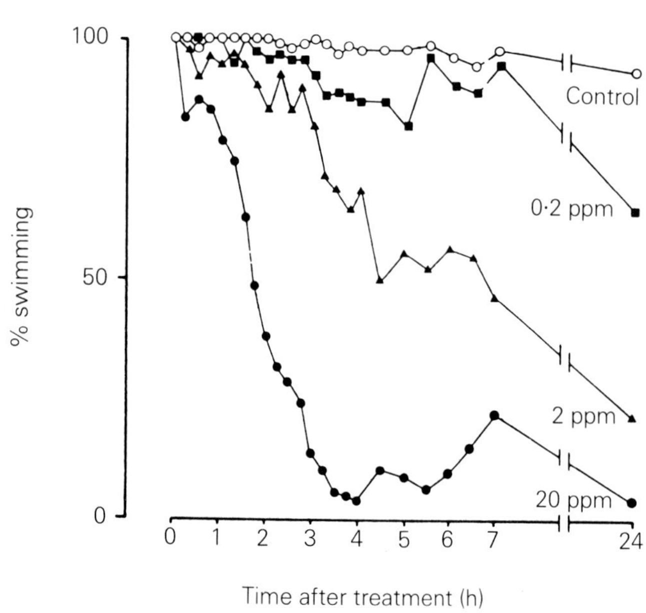

Figure 5. Effects of cyanatryn on swimming activity in Daphnia pulex *(source: Scorgie & Cooke 1979)*

The problem is more one of minimizing the indirect side-effects of herbicide treatments, such as deoxygenation. One solution may be to limit the use of herbicides to partial treatment of a fishery. This limited use is relatively easy with emergent treatments, where the spray can be applied only to those areas of the reed bed where it is wanted. Dalapon has been used in this way in Nature Reserves, such as the Ouse Wash Reserve of the Royal Society for the Protection of Birds, to clear out areas of open water in dense reedswamp vegetation for waterfowl. The gel formulation of diquat-alginate has also proved suitable for this purpose, on submerged plants, in standing and flowing waters (Barrett & Logan 1982; Barrett & Murphy 1982). Other options to minimize the environmental impact of treatment include treating at less sensitive times of year: glyphosate, for example, may be applied to branched bur-reed (*Sparganium erectum*) in early autumn, causing a very natural-looking dieback of the plants and preventing (because it kills the rhizome system) regrowth in thc following season (Eaton *et al.* 1981).

3.3.5 Biological weed control

Direct biological control of aquatic plant growth using, in particular, herbivorous fish such as the grass carp (*Ctenopharyngodon idella*), a species native to northern China, has been under investigation in Britain since the 1960s. One reason for the slow progress of the research has been that a number of problems, potential or real, required detailed investigation before any release of an organism with the potential to cause gross habitat modification could be contemplated. Primary among these considerations was the question of whether the fish would be able to breed in British waters but this now appears unlikely (Stott *et al.* 1971). Populations of the fish, artificially bred, can therefore be used and controlled fairly easily, but are not a panacea solution to weed problems in fishery waters.

Difficulties include the following.

i. The grass carp is a selective grazer (Fowler & Robson 1978), preferring tender foliage, such as that of Canadian pondweed, to tougher macrophyte species such as fan-leaved water-crowfoot (*Ranunculus circinatus*), which survived even high grazing pressure by grass carp in Dutch polder waterways. This fact leaves open the possibility of shifts of weed species as a result of the use of grass carp, with one weed problem being substituted by another less susceptible to the new control regime. Also likely are effects on epiphytic invertebrate communities (van der Zweede 1982).

ii. Because the grass carp has a relatively short gut, only some 10% of ingested vegetation is digested. There is, therefore, a tendency to recirculate nutrients, hitherto locked up in macrophytes, into the waterbody, with the likely consequence of stimulating phytoplankton blooms (van der Zweerde *et al.* 1978).

iii. The fish feeds poorly, or not at all, at the low temperatures (10°C) typical of conditions during spring in British fresh waters, when plant growth is rapidly increasing. If vegetative growth outstrips the grazing ability of the fish at this crucial part of the season, a serious weed problem can ensue, despite the later-season grazing and reduction of plant growth (Mugridge *et al.* 1982).
iv. Fisheries managers have, in general, an ambivalent attitude to the balance between the potential benefits to be gained from biological weed control, and the potential problems associated with the deliberate introductions of an alien fish species.

These potential problems notwithstanding, grass carp are an attractive proposition for plant management in many fisheries, and by 1985 were being introduced on a moderately wide-scale experimental basis by several English Water Authorities, and also to a few waters in Scotland.

It should perhaps be mentioned that other, native, fish species may, through various mechanisms, cause a lesser degree of modification of plant communities in fresh waters (Andersson *et al.* 1978). Cyprinid species such as roach may graze the shoots of submerged plants (Prejs 1978). More important are the effects of bottom-feeders, such as common carp (*Cyprinus carpio*), which stir up sediments and increase organic loadings via their excrement, ultimately reducing light penetration of the water column, with adverse effects on submerged plant growth (Lamarra 1975). At moderately high population densities, carp may significantly reduce the growth of submerged macrophytes (King & Hunt 1967).

At present, besides the use of herbivorous fish, only one other biological approach to aquatic weed control is used in British fishery waters. Shading by trees or bushes is a long-established technique which has fallen into some disfavour as a management approach over the last 50 years. Recently, however, shading has been the subject of renewed interest. For example, the relationship between shade density and orientation, and macrophyte growth in shaded rivers or streams has been quantified (Dawson & Kern-Hansen 1979). As a procedure of 'biotechnical engineering', this approach appears to be regaining some favour amongst river management organizations, with potential improvements to the overall status of the watercourse as a wildlife habitat. However, it is important to maintain a long-term cycle of planting and clearance of the shade vegetation, so as to adjust the light regime within the stream to permit a moderate degree of macrophyte growth. A heavily shaded system tends to support an impoverished aquatic ecosystem and lower fish population (White 1975). The absence of macrophyte oxygen production within the water column, coupled with the high oxygen demand arising from the decay of allochthonous leaf litter, may lead to local deoxygenation problems (Dawson 1976b).

4 Case studies of habitat modification procedures undertaken for fisheries management purposes

Two contrasting forms of plant management aimed at improving the fisheries status of a freshwater system are examined briefly in the following studies.

4.1 Vegetation management to increase macrophyte cover: the lower Welsh Dee, north Wales

The Welsh Dee supports important recreational coarse and game fisheries. In the early 1970s, the Welsh Water Authority received complaints about a decline in the quality of coarse angling in the lower reaches of the river near Chester. Following research by O'Hara (1976) who postulated that a lack of aquatic plants, possibly due to the effects of the large number of boats on the river, was affecting the recruitment of coarse fish (especially roach) the Welsh Water Authority commissioned a research project to examine the ecology of the lower Welsh Dee and to investigate possible methods to improve the fishery. Some of the findings of this work are reported in Pearce (1983a,b) and Pearce and O'Hara (1984).

Amongst the studies undertaken were 2 experimental projects to establish the feasibility of protecting, and artificially increasing, aquatic vegetation in the river for the spawning and recruitment of coarse fish. These projects involved studying the effect of protecting existing floating-leaved aquatic plant growth (yellow water-lily (*Nuphar lutea*)) from boats' propellers and examining the growth of artificially introduced emergent, floating-leaved and submerged aquatic plant species (Plate 10). Beneficial effects resulted from these treatments. At peak biomass, greater numbers of surface leaves, flowers, roach fry and invertebrates were recorded from lily beds protected from boats, compared with control areas. Spectacular growths of the artificially introduced emergent, floating-leaved, and submerged aquatic plant species were observed, especially at sites where boats and cattle were excluded, but were reversed when exclusion devices were removed. Further trials are now being conducted. In river lengths with little aquatic vegetation due to extensive tree shading, short lengths of trees are also being removed and emergent 'reed' species and lilies are being introduced to increase fish habitat and the quality of the river environment generally.

4.2 Vegetation management to reduce macrophyte biomass on a selective basis: The Laird's Loch, Dundee

The Laird's Loch is a small (6.3 ha) upland loch in the Sidlaw Hills near Dundee, Scotland. It is a designated Site of Special Scientific Interest and supports a trout fishery. Excessive submerged macrophyte growth, mainly thread-leaved water-crowfoot (*Ranunculus trichophyllus*), had caused problems for boat fly fishing here for several years up to 1984 (Figure 6). Attempts

Plate 10. Protected plantings of emergent and floating-leaved vegetation (fenced to prevent cattle grazing and boat access) on a vegetation-denuded bank of the River Dee, Chester (Photograph K J Murphy)

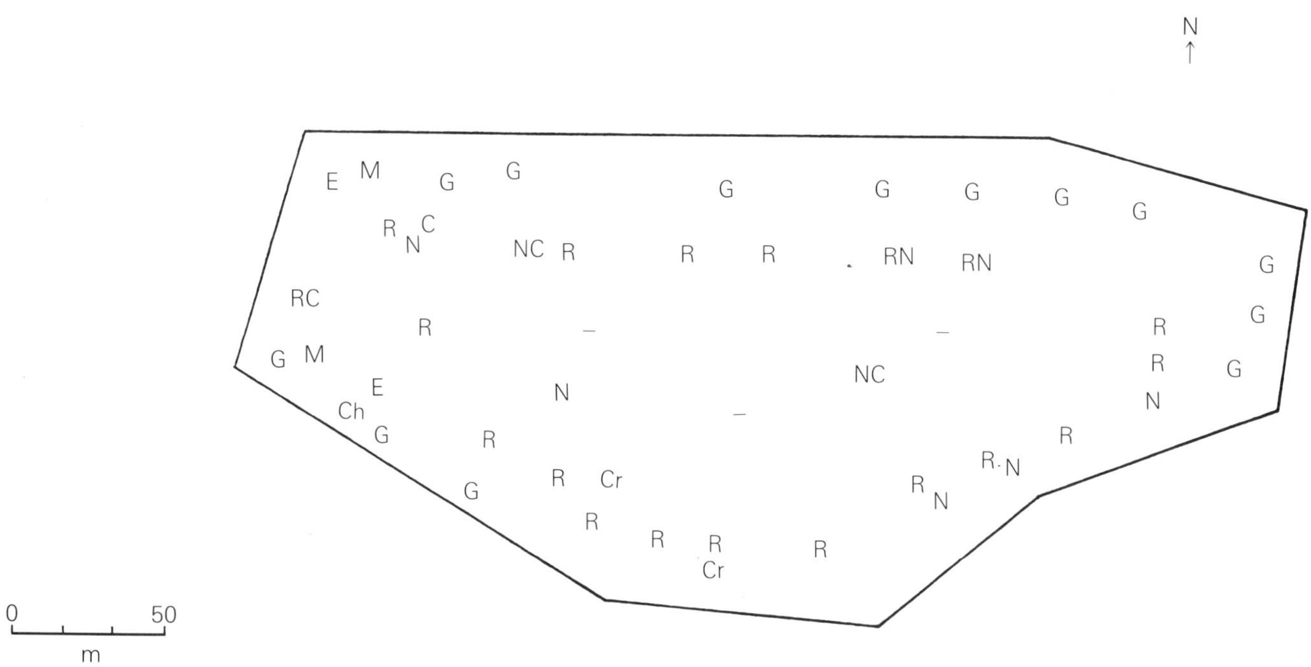

*Figure 6. Submerged vegetation map of Laird's Loch, Dundee (June 1984). Experimental plots were located in the zones dominated by thread-leaved water crowfoot (*Ranunculus trichophyllus*), on the north and south sides of the loch (reproduced by permission of the Nature Conservancy Council and Ms K Anderson)*

R Ranunculus trichophyllus
G Potamogeton x nitens
Cr Potamogeton crispus
M Myriophyllum alterniflorum
E Elodea canadensis
C Callitriche hermaphroditica
N Nitella opaca
Ch Chara globularis

to control the growth with liquid formulation diquat during the 1970s, by the owners, met with little success. In June 1984, an experimental application of diquat-alginate at 1.0 mg/litre was made at the request of the Nature Conservancy Council, with the aim of reducing weed growth in 2 small plots (each approximately 100 m^2). The results (Figure 7) showed that, whilst growth increased steadily in the untreated plots during the summer, there was a net decrease in biomass in the treated plots. By the end of the season, there was a significantly higher biomass of macrophytes present in the untreated than in the diquat-treated plots. Weed growth had reached the surface over a wide area of the loch, but was only a few centimetres high within the treated plots. There was no increase in phytoplankton biomass, and no significant change in water chemistry in treated plots, compared with the control plots on equivalent sampling dates, although a slight reduction (to approximately 11 mg/litre) in daytime dissolved oxygen occurred in July at one of the treated plots. Monitoring of plankton Cladoceran populations in the experimental plots revealed no major undesirable effects on this non-target group of organisms (B R S Morrison, DAFS Freshwater Fisheries Laboratory, personal communication), which is known to possess fairly low margins of tolerance to diquat (Crosby & Tucker 1966). The successful outcome of this trial demonstrated the potential of diquat-alginate for weed control for fisheries management purposes, without causing undue habitat side-effects.

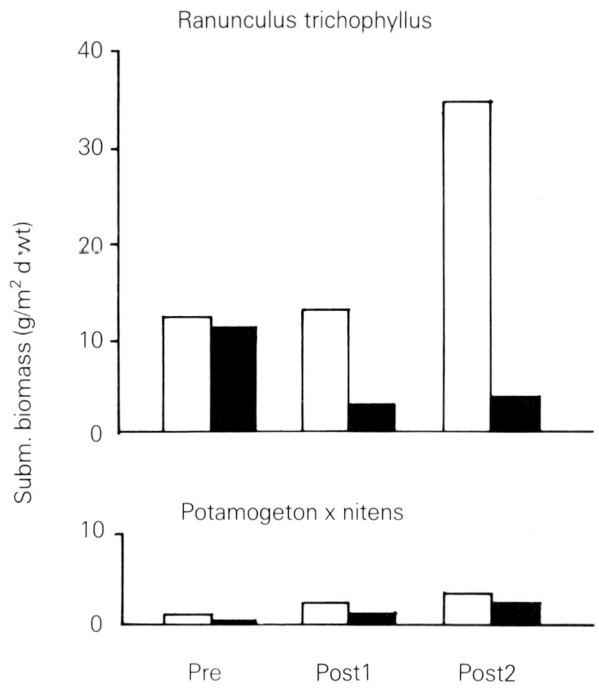

Figure 7. Effects of diquat-alginate treatment on dry weight biomass of 2 dominant macrophyte species in Laird's Loch during summer 1984 (■ treated, □ untreated; PRE = 7 June 1984; POST1 = 20 July 1984; POST2 = 7 September 1984). Treatments significant at P=0.05: analysis of variance, random-block model. Least significant difference between pairs of means: 24.7 (data reproduced by permission of Ms K Anderson)

5 Conclusions

Habitat conditions in an aquatic system essentially define not only the abundance and species of fish present, but also the range of wildlife species which are dependent, at least in part, on the waterbody for their survival.

Perhaps only slightly overstating the case, Lewis and Williams wrote, in 1984, that 'the plant community is the most valuable element of the river ecosystem, on which every other living thing depends: the more varied its structure and composition, the greater the diversity of other wildlife it can support'.

However, they also recognized that '. . . some plants, at particular locations or times of year can be a nuisance to man . . . the river manager therefore needs to know plants and their growth habits, and recognise potential nuisance'.

Most anglers and fishery managers recognize the need to strike a balance in managing the vegetation of the waters they fish or manage. Too much vegetation, or the wrong type of plants can be a problem, but a source of equal concern is a lack of vegetation. Appropriate and sensitive care for, and management of, the plant growth, and chemical and physical conditions of the freshwater habitat are in the long run to the benefit of the angler, as much as the wildlife of the system.

6 Summary

There are 2 principal aspects of habitat modification associated with angling that may affect the environment. The magnitude of the direct impacts on the water's edge and open water habitats is related quite closely to the type of waterbody and form of angling involved, and may vary in importance between sites and different times of year. The timing of the coarse fishing close season probably reduces the impacts of anglers on habitats during a critical part of the plant growth season. Habitat modification may also be produced by fisheries management procedures. Such management may be aimed at improving anglers' access to a fishery, or altering the freshwater environment to manipulate fish population size, biomass or structure, thereby improving the anglers' chance of catching fish. Typical modifications involve altering physical or chemical conditions within the waterbody (eg by dredging, installation of flow-modifying structures, liming, manipulation of nutrient status). Management of aquatic plant growth, with associated effects on invertebrates and other organisms important in fish food chains, is a long-established form of habitat modification. Weed cutting, herbicides or biological weed control methods may have profound effects on the fish populations and other wildlife of a freshwater habitat. Conversely, management aimed at increasing plant growth in or around waterbodies may also have potentially important consequences for the breeding success and survival of fish and other organisms. Case studies of the habitat modification implications of fishery management procedures are given for a lowland river and an upland loch.

7 References

Andersson, G., Berggren, H., Cronberg, G. & Gelin, C. 1978. Effects of planktivorous and benthivorous fish on organisms and water chemistry in eutrophic lakes. *Hydrobiologia*, **59**, 9–15.

Barrett, P. R. F. & Logan, P. 1982. The localised control of submerged aquatic weeds in lakes with diquat-alginate. *Proc. Symp. Aquat. Weeds, 6th, 1982*, 193–199.

Barrett, P. R. F. & Murphy, K. J. 1982. The use of diquat-alginate for weed control in flowing waters. *Proc. Symp. Aquat. Weeds, 6th, 1982*, 200–208

Bell, D., Odin, N., Austin, L., Hayhow, S., Jones, A., Strang, A. & Torres, E. 1984. *The impact of anglers on wildlife and site amenity.* (Report.) Cardiff: University of Wales Institute of Science & Technology.

Birch, E. 1976. Trout stream weeds: how to cut and how not to, for fish and fly. *Salm. Trout Mag.*, **206**, 49–53.

Bonham, A. J. 1980. *Bank protection using emergent plants against boat wash in rivers and canals.* (Report no. IT 206.) Wallingford, Oxon: Hydraulics Research Station.

Bouquet, H. G. J. 1978. Fisheries and waterweeds. *Proc. Symp. Aquat. Weeds, 5th, 1978*, 79–82.

Bregazzi, P. R., Domaniewski J. C. J. & Jones, J. G. W. 1984. *Carp farming project: final report.* Reading: University of Reading, Dept of Agriculture and Horticulture.

British Waterways Board. 1980. *Recreation on inland waterways.* 2nd ed. London: British Waterways Board.

British Waterways Board. 1981. *Vegetation control manual.* London: British Waterways Board.

British Waterways Board. 1982. *Reinforced vegetative bank protection: 1982 trials report.* Gloucester: British Waterways Board.

Brooker, M. P. 1974. The risk of deoxygenation of water in herbicide application for aquatic weed control. *J. Inst. Wat. Engrs*, **28**, 206–210.

Brooker, M. P. & Edwards, R. W. 1975. Review paper: aquatic herbicides and the control of water weeds. *Wat. Res.*, **9**, 1–15.

Brooker, M. P., Morris, D. L. & Hemsworth, R. J. 1977. Mass mortalities of adult salmon (*Salmo salar*) in the R. Wye, 1976. *J. appl. Ecol.*, **14**, 409–417.

Brown, D. J. A. 1982. The effect of pH and calcium on fish and fisheries. *Water Air Soil Pollut.*, **18**, 343–352.

Burns, J. C., Coy, J. S., Tervet, D. J., Harriman, R., Morrison, B. R. S. & Quine, C. P. 1984. The Loch Dee project: a study of the ecological effects of acid precipitation and forest management on an upland catchment in southwest Scotland. 1. Preliminary investigations. *Fish. Manage.*, **15**, 145–167.

Carpenter S. R. & Greenlee, J. K. 1981. Lake deoxygenation after herbicide use: a simulation model analysis. *Aquat. Bot.*, **11**, 173–186.

Casey, H. & Westlake, D. F. 1974. Growth and nutrient relationships of macrophytes in Sydling Water, a small unpolluted chalk stream. *Proc. Symp. Aquat. Weeds, 4th*, 69–76

Cooley, T. N., Dooris, P. M. & Martin, D. F. 1980. Aeration as a tool to improve water quality and reduce the growth of *Hydrilla. Wat. Res.*, **14**, 485–489.

Crosby, D. G. & Tucker, R. K. 1966. Toxicity of aquatic herbicides to *Daphnia magna. Science, N.Y.*, **154**, 289–291.

Dartington Amenity Research Trust. 1973. *Rickmansworth gravel pits.* (Sports Council study no. 4.) London: The Sports Council.

Dawson, F. H. 1976a. The annual production of the aquatic macrophyte *Ranunculus penicillatus* var. *calcareus* (R. W. Butcher) C. D. K. Cook. *Aquat. Bot.*, **3**, 51–73.

Dawson, F. H. 1976b. Organic contribution of stream edge forest litter fall to chalk stream ecosystems. *Oikos*, **27**, 13–18.

Dawson, F. H. & Kern-Hansen, U. 1979. The effect of natural and artificial shade on the macrophytes of lowland streams and the use of shade as a management technique. *Int. Revue ges. Hydrobiol.*, **64**, 437–455.

Dobolyi, E. & Herodek, S. 1980. On the mechanism reducing the phosphorus concentration in the water of Lake Balaton. *Int. Revue ges. Hydrobiol.*, **65**, 339–343.

Dvorak, J. & Best, E. P. H. 1982. Macroinvertebrate communities associated with the macrophytes of Lake Vechten: structural and functional relationships. *Hydrobiologia*, **95**, 115–126.

Eaton, J. W., Murphy, K. J. & Hyde, T. M. 1981. Comparative trials of herbicidal and mechanical control of aquatic weeds in canals. *Proc. AAB Conf. Aquatic Weeds & their Control, 1981*, 105–116.

Edwards, R. W. & Bell, D. 1985. Fishing in troubled waters. *New Scient.*, **105**, 19–21.

Edwards, R. W. & Bell, D. 1987. The impact of angling on wildlife. *Proc. Br. Freshwat. Fish. Conf., 4th, 1985.* In press.

Fowler, M. C. & Robson, T. O. 1978. The effects of the food preferences and stocking rate of grass carp (*Ctenopharyngodon idella* Val.) on mixed plant communities. *Aquat. Bot.*, **5**, 261–276.

George, M. 1976. Mechanical methods of weed control in watercourses – an ecologist's view. *Proc. Symp. Aquat. Herbicides*, 91–99. (British Crop Protection Council monograph no. 16.)

Hall, R. 1987. Leisure on inland waterways. *Proc. Br. Freshwat. Fish. Conf., 4th, 1985.* In press.

Haller, W. T., Shireman, J. V. & Durant, D. F. 1980. Fish harvest resulting from mechanical control of *Hydrilla. Trans Am. Fish. Soc.*, **109**, 517–520.

Harbott, B. J. & Rey, C. J. 1981. The implications of long-term aquatic herbicide application: problems associated with environmental impact assessment. *Proc. AAB Conf. Aquat. Weeds & their Control, 1981*, 219–231.

Harriman, R. & Morrison, B. R. S. 1982. Ecology of streams draining forested and non-forested catchments in an area of central Scotland subject to acid precipitation. *Hydrobiologia*, **88**, 251–263.

Hultberg, H. & Andersson, I. B. 1982. Liming of acidic lakes: induced long-changes. *Water Air Soil Pollut.*, **18**, 311–331.

Hunt, R. C. 1968. Effects of habitat alteration on production, standing crop and yield of brook trout in Lawrence Creek, Wisconsin. *Sport Fishery Abstr.*, **13**, 2, 9899.

Inland Waterways Amenity Advisory Council. 1975. *Angling on the British Waterways Board system.* London: IWAAC.

Inland Waterways Amenity Advisory Council. 1983. *Waterways ecology and the design of recreational craft.* London: IWAAC.

Irwin, W. H., Symons, J. M. & Robuck, G. G. 1966. Impoundment stratification by mechanical pumping. *J. sanit. Engng Div. Am. Soc. civ. Engrs*, **92**, (SA6) 21–40.

International Union for the Conservation of Nature and Natural Resources. 1967. *Towards a new relationship of man and nature in temperate land. Part 1. Ecological impact of recreation and tourism upon temperate environments.* (IUCN publ. new ser. no. 7.)

Keast, A. 1984. The introduced aquatic *Myriophyllum spicatum* as habitat for fish and their invertebrate prey. *Can. J. Zool.*, **62**, 1289–1303.

Kelcey, J. G. 1978. Techniques No. 31. Creative ecology Part 2. Selected aquatic habitats. *Landscape Des.*, **121**, 36–38.

Kelcey, J. G. 1981. Weed control in amenity lakes. *Proc. AAB Conf. Aquat. Weeds & their Control, 1981*, 15–31.

Kern-Hansen, U. 1978. Drift of *Gammarus pulex* L. in relation to macrophyte cutting in four small Danish lowland streams. *Verh. int. Verein. theor. angew. Limnol.*, **20**, 1440–1445.

King, D. R. & Hunt, G. S. 1967. Effect of carp on vegetation in a Lake Erie marsh. *J. Wildl. Mgmt*, **31**, 181–188.

Kirby G. E. & Shell, F. W. 1976. Karmex herbicide with fertilisation and aeration to control filamentous algae in hatchery ponds. *Proc. Ann. Conf. S.E. Ass. Game Fish Comm., 1975*, **19**, 279–281.

Laing, R. L. 1979. The use of multiple inversion and 'clean-Flo Lake Cleanser' in controlling aquatic plants. *J. aquat. Plant Manage.*, **17**, 33–38.

Lamarra, V. A. 1975. Digestive activities of carp as a major contributor to the nutrient loading of lakes. *Verh. int. Verein. theor. angew. Limnol.*, **19**, 2461–2468.

Lange, L. de & van Zon, J. C. J. 1978. Evaluation of the botanical response of different methods of aquatic weed control, based on the structure and floristic composition of the macrophyte vegetation. *Proc. Symp. Aquat. Weeds, 5th, 1978*, 279–286.

Lewis, G. & Williams, G. 1984. *Rivers and wildlife handbook.* Lincoln: Royal Society for the Protection of Birds/Royal Society for Nature Conservation.

Liddle, M. J. & Scorgie, H. R. A. 1980. The effects of recreation on freshwater plants and animals: a review. *Biol. Conserv.*, **17**, 183–206.

Lindmark G. K. 1984. Acidified lakes: ecosystem response following sediment treatment with sodium carbonate. *Verh. int. Verein. theor. angew. Limnol.*, **22**, 772–779.

Lorenzen, M. W. 1977a. Aeration circulation keeps algal blooms in check. 1. Mixing and destratification techniques. *Wat. Wastes Eng.*, **14**, 69–74.

Lorenzen, M. W. 1977b. Aeration circulation keeps algal blooms in check. 2. Diffuser design and compressor selection. *Wat. Wastes Eng.*, **14**, 88–92.

Maitland, P. S. 1972. *Key to British freshwater fishes.* (Sci. Publ. no. 27.) Ambleside: Freshwater Biological Association.

Marshall, E. J. P. & Westlake, D. F. 1978. Recent studies on the role of aquatic macrophytes in their ecosystem. *Proc. Symp. Aquat. Weeds, 5th, 1978*, 43–51.

Martin, D. F., Doig, M. T. & Millard, K. 1971. Potential control of Florida *Elodea* by nutrient-control agents. *Hyacinth Control J.*, **9**, 36–39.

Mills, C. A. 1981. The spawning of roach *Rutilus rutilus* L. in a chalk stream. *Fish Manage.*, **12**, 49–54.

Mills, D. H. 1979. The impact of angling on the environment. *Proc. Inst. Fish. Manage., 10th Study Course, Nottingham University, 1979*, 5–16.

Ministry of Agriculture, Fisheries & Food. 1979. *Guidelines for the use of herbicides in or near watercourses and lakes.* (Booklet no. 2078.) London: MAFF.

Mortensen, E. 1977. Density-dependent mortality of trout fry (*Salmo trutta* L.) and its relationship to the management of small streams. *J. Fish Biol.*, **11**, 613–617.

Mugridge, R. E. R., Buckley, B. R., Fowler, M. C. & Stallybrass, H. G. 1982. An evaluation of the use of grass carp (*Ctenopharyngodon idella* Val.) for controlling aquatic plant growth in a canal in southern England. *Proc. int. Symp. Herb. Fish, 2nd, 1982*, 8–16.

Murphy, K. J. 1982. The use of methylthio-triazine herbicides in freshwater systems: a review. *Proc. Symp. Aquat. Weeds, 6th, 1982*, 263–277.

Murphy, K. J. & Eaton, J. W. 1981. Water plants, boat traffic and angling in navigable canals. *Proc. Br. Freshwat. Fish. Conf., 2nd, 1981*, 173–187.

Murphy, K. J. & Eaton, J. W. 1983. Effects of pleasure-boat traffic on macrophyte growth in canals. *J. appl. Ecol.*, **20**, 713–729.

Murphy, K. J., Bradshaw, A. D. & Eaton, J. W. 1980. *Plants for the protection of canal banks.* Report to British Waterways Board. Liverpool: University of Liverpool, Dept of Botany.

Murphy, K. J., Eaton, J. W. & Hyde, T. M. 1982. The management of aquatic plants in a navigable canal system used for amenity and recreation. *Proc. Symp. Aquat. Weeds, 6th, 1982*, 141–151.

National Opinion Polls. 1980. *National angling survey 1980.* Summary report. London: National Opinion Polls.

Newbold, C. & Palmer, M. 1979. *Aquatic plants as indicators of the trophic status of standing waters.* (CST notes no. 18.) Huntingdon: Nature Conservancy Council.

Northcott, D. 1979. The importance of aquatic macrophytes in the provision of crustacean zooplankton food for young roach. *Proc. Br. Freshwat. Fish. Conf., 1st, 1979*, 123–134.

O'Grady, K. T. & Spillett, P. B. 1981. The production of carp (*Cyprinus carpio*) in disused filter beds. *Proc. Br. Freshwat. Fish. Conf., 2nd, 1981*, 17–28.

Ogwada, R. A., Reddy, K. R. & Graetz, D. A. 1984. Effects of aeration and temperature on nutrient regeneration from selected aquatic macrophytes. *J. environ. Qual.*, **13**, 239–242.

O'Hara, K. 1976. *An ecological study of fishes, with particular reference to freshwater species, in the Welsh Dee at Chester.* Ph.D. thesis, University of Liverpool.

Parker, R. V., Stone, P. B., Buchanan, C. C. & Steimle, R. 1974. *How to build artificial reefs.* (Fishery Facts 10.) NOAD, U.S. National Marine Fisheries Service.

Pearce, H. G. 1983a. Management strategies for British coarse fisheries: the lower Welsh Dee, a case study. *Proc. Br. Freshwat. Fish. Conf., 3rd, 1983*, 262–273.

Pearce, H. G. 1983b. Coarse fish stocking of British rivers: a case study of the lower Welsh Dee. *Proc. Inst. Fish. Manage., 14th Study Course, City University, London, 1983*, 209–220.

Pearce, H. G. & O'Hara K. 1984. *The management and ecology of the Welsh Dee – an annotated bibliography.* Stevenage: Water Research Centre.

Pearson, R. G. & Jones, N. V. 1975. The effects of dredging operations on the benthic community of a chalk stream. *Biol. Conserv.*, **8**, 273–278.

Prejs, A. 1978. Lake macrophytes as the food of roach (*Rutilus rutilus* L.) and rudd (*Scardinius erythrophthalamus* L.) II. Daily intake of macrophyte food in relation to body size of fish. *Ekol. polsk.*, **26**, 537–553.

Price, H. 1981. A review of current mechanical methods. *Proc. AAB Conf. Aquat. Weeds & their Control, 1981*, 77–86.

Rabe, F. W. & Gibson, F. 1984. The effect of macrophyte removal on the distribution of selected invertebrates in a littoral environment. *J. Freshwat. Ecol.*, **2**, 359–371.

Rees, J. & Tivy, J. 1978. Recreational impact on Scottish lochshore wetlands. *J. Biogeogr.*, **5**, 93–108.

Salanki, J. & Biro, P., eds. 1979. *Human impacts on life in freshwaters.* Budapest: Akademiai Kiado.

Satchell, J. E. 1976. *The effects of recreation on natural landscapes.* (Nature and Environment Series, no. 11.) Brussels: Council of Europe.

Savino, J. P. & Stein, R. H. 1982. Predator-prey interaction between largemouth bass and bluegills, as influenced by simulated submerged vegetation. *Trans Am. Fish. Soc.*, **111**, 255–266.

Schaeperclaus, W. 1933. *Textbook of pond culture.* Translated by F. Hund, Stanford University. Berlin: Paul Parey.

Scheffer, M., Achterberg, A. A. & Beltman, B. 1984. Distribution of macroinvertebrates in a ditch in relation to the vegetation. *Freshwater Biol.*, **14**, 367–370.

Scorgie, H. R. A. & Cooke, A. S. 1979. Effects of the triazine herbicide cyanatryn on aquatic animals. *Bull. environ. Contam. Toxicol.*, **22**, 135–142.

Shireman, J. V., Haller, W. T., Colle, D. E. & Durant, D. F. 1983. Effects of aquatic macrophytes on native sportfish populations in Florida. *Proc. int. Symp. Aquat. Macrophytes, Nijmegen, 1983*, 208–214.

Soulsby, P. G. 1974. The effect of a heavy cut on the subsequent growth of aquatic plants in a Hampshire chalk stream. *J. Inst. Fish. Mgmt*, **5**, 49–53.

Stabler, M. J. & Ash, S. E. 1977. *The amenity demand for inland waterways. Angling. A preliminary report.* Reading: Reading University, Amenity Waterways Study Unit.

Stott, B., Cross, D. G., Iszard, R. E. & Robson, T. O. 1971. Recent work on the grass carp in the UK from the standpoint of its economics in controlling submerged aquatic plants. *Proc. int. Symp. Aquat. Weeds, 3rd, 1971*, 105–116.

Sukopp, H. 1971. Effects of man, especially recreational activities, on littoral macrophytes. *Hydrobiologia*, **12,** 331–340.

Swales, S. 1982. Impacts of weed-cutting on fisheries: an experimental study in a lowland river. *Fish. Manage.*, **13**, 125–138.

Swales, S. & O'Hara, K. 1983. A short-term study of the effects of a habitat improvement programme on the distribution and abundance of fish stocks in a small lowland river in Shropshire. *Fish. Manage.*, **14**, 135–144.

Thomas, E. A. 1978. Mass growth of algae and macrophytes in streams: method, cause, and prevention. *Verh. int. Verein. theor. angew. Limnol.*, **20**, 1796–1799.

Tivy, J. & O'Hare, G. 1981. *Human impact on the ecosystem.* Edinburgh: Oliver & Boyd.

Tooby, T. E. 1978. A scheme for the evaluation of hazards to non-target aquatic organisms from the use of chemicals. *Proc. int. Symp. Aquat. Weeds, 5th, 1978*, 287–294.

Tooby, T. E. & Spencer-Jones, D. H. 1978. The fate of the aquatic herbicide dichlobenil in hydrosoil, water and roach (*Rutilus rutilus* L.) following treatment of three areas of a lake. *Proc. int. Symp. Aquat. Weeds, 5th, 1978*, 323–331.

Vollenweider, R. A. 1968. *Scientific fundamentals of the eutrophication of lakes and flowing waters, with particular reference to nitrogen and phosphorus as factors in eutrophication.* (Technical Report Water Management & Research no. DAS/CSI/6.27.)

Wade, P. M. & Edwards, R. W. 1980. The effect of channel maintenance on the aquatic macrophytes of the drainage channels of the Monmouthshire Levels, south Wales, 1840–1976. *Aquat. Bot.*, **8**, 307–322.

Wade, P. M. 1982. The long-term effects of herbicide treatments on aquatic weed communities. *Proc. int. Symp. Aquat. Weeds, 6th, 1982*, 278–285.

White, R. J. 1975. Trout population responses to streamflow fluctuation and habitat management in Big Roch-a-Cri Creek, Wisconsin. *Verh. int. Verein. theor. angew. Limnol.*, **19**, 2469–2477.

Wingfield, G. I. & Bebb, J. M. 1982. Simulation of deoxygenation of water containing vascular plants following treatment with terbutryne. *Proc int. Symp. Aquat. Weeds, 6th, 1982*, 255–262.

Wolny, P. 1967. Fertilization of warm-water fish ponds in Europe. *FAO Fish. Rep.*, **44**, 64–81. Rome: FAO.

Wright, D. M. 1973. *The fly-fisher's plants: their value in trout waters.* Newton Abbot: David & Charles.

Zweerde W. van der, Hoogers, B. J. & Zon, J. C. J. van. 1978. Effects of grass carp on microflora and -fauna, macroflora, macroinvertebrates and chemical properties in the water. *Proc. int. Symp. Aquat. Weeds, 5th, 1978*, 158–164.

Zweerde W. van der. 1982. The influence of grass carp (*Ctenopharyngodon idella* Val.) on macroinvertebrates in some experimental ditches in the Netherlands. *Proc. int. Symp. Herbivorous Fish, 2nd, 1982*, 158–164.

Use and effects of piscicides

B R S MORRISON
DAFS, Freshwater Fisheries Laboratory, Pitlochry

1 Introduction

In fisheries management it is often necessary to control the numbers of an undesirable species or obtain samples of a population for study. This can often be done using nets or traps and, in rivers and streams, by electrofishing. Where complete elimination of a fish population is planned, these methods may not be suitable, either because they are selective and capture only certain sizes of fish, or the time and manpower required make them impracticable. In these situations, the use of a piscicide may be considered, provided side-effects on other biota are acceptable.

During the past 50 years, many chemicals have been used as piscicides, including inorganic substances such as ammonia, chlorine, copper sulphate, sodium hydroxide and sodium cyanide. Amongst the organic chemicals, man-made materials like malathion, dichlorvos and endrin, originally developed as insecticides, have all been used to kill fish (Lennon *et al.* 1970). Although a few of these substances were found to be useful in some circumstances (eg malathion could be used selectively because it is less toxic to some fish species than others), many proved to be highly toxic not only to fish but also to birds and mammals. An aerial application giving 0.52 mg/litre of toxaphene in the water killed all the fish in a lake in Nebraska, and almost 10% of the gulls (*Larus* spp.) and grebes (*Podiceps* and *Tachybaptus* spp.), 29% of gadwall (*Anas strepera*) and 33% of the mallard (*Anas platyrhynchos*) (D B McCarraher & J L Dean, unpublished data). A further disadvantage of these pesticides is their persistence in the water. A lake in British Columbia treated with toxaphene was found to be still toxic 9 months later (Stringer & McMinn 1958), and polychlorpinene used to treat about 250 lakes in the Soviet Union persisted in some for one and half years (Lennon *et al.* 1970).

The ideal piscicide should have a low toxicity for other forms of animal life; it should degrade rapidly in water, and be relatively unaffected by temperature, pH and alkalinity. Several of these qualities are found in piscicides derived from roots, leaves or seeds of plants which have been used for centuries for this purpose in South America, Africa and Asia, either to collect fish for food (Leonard 1939; Krumholz 1948) or to remove unwanted species from culture ponds (Tang 1961). Of these naturally occurring piscicides, the saponins, water-soluble glycosides, are perhaps the most common, being present in at least 75 families of plants, including azaleas, rhododendrons, camellias and heaths. In the Soviet Union, research on saponins derived from sugar beet showed that they were toxic to fish at concentrations of 0.2 mg/litre and broke down in water within 10 days (Bizyaev *et al.* 1965). In this paper, the use and effects of rotenone, probably the most widely used piscicide, and of antimycin A, a recently developed piscicide, will be considered.

2 The piscicides

Rotenone is the toxic principle in at least 6 genera of leguminous plants. When dried and powdered, the roots of *Derris*, the genus most commonly used in Asia, contain 4–5% by weight of rotenone. In other genera, eg *Lonchocarpus* and *Tephrosia*, used in North and South America, rotenone content varies from 2% to 5% (Pintler & Johnson 1958). The powdered root is relatively insoluble and in the commercial development of rotenone as a piscicide from about 1930 onward the manufacture of wettable powders and emulsions made application easier. Some liquid formulations contain 5% rotenone, others 2.5%, with synergists to enhance the toxic effect. Unfortunately, the solvents in some of these formulations repel fish and care has to be taken to prevent fish escaping from the treatment area.

Rotenone acts by constricting the blood vessels of the gills, thus hindering respiration (Hamilton 1941). Biochemically it has been shown to inhibit the transfer of electrons along the respiratory pathway (Oberg 1967). In Scottish lochs, concentrations of 0.05 mg/litre have been found by staff at the Freshwater Fisheries Laboratory, Pitlochry, to be effective against pike (*Esox lucius*), perch (*Perca fluviatilis*), eels (*Anguilla anguilla*) and trout (*Salmo trutta*). In streams, a concentration of about 0.5 mg/litre for 30 min is used in controlling trout populations. Toxicity to birds and mammals is much less than for fish. The oral LD_{50} for mallard and pheasants (*Phasianus colchicus*) is over 1 g/kg, for rabbits (*Oryctolagus cuniculus*) 1.5 g/kg and for dogs 3 g/kg (Lennon *et al.* 1970). In experiments with southern leopard frog tadpoles (*Rana sphenocephala*), however, Chandler (1982) recorded LD_{50} values of 0.83 mg/litre (for a period of 1 h), close to the concentrations used in stream treatments.

The toxicity of rotenone decreases on exposure to light, heat, oxygen, alkalinity and turbidity (Almquist 1959). A synergized product held in drums at the Freshwater Fisheries Laboratory was still toxic 5 years after delivery. In ponds where there is little organic matter, rotenone can be detoxified by potassium permanganate or chlorine (chlorinated lime) applied at the same concentration as the rotenone, but, because both detoxifying agents have a strong affinity for organic matter, higher concentrations will be required where deep silt or dense plant life is present (Jackson

1967). Concentrations of potassium permanganate from 2.0 to 2.5 mg/litre successfully neutralized 0.05 mg/litre rotenone in experimental ponds (Lawrence 1956). In still water, rotenone degrades naturally over a period of 2 weeks or more, depending on temperature. When sampling fish in a stream, Mahon and Balon (1980) used 6–7 mg/litre potassium permanganate to detoxify 1.0–1.5 mg/litre Pro-noxfish (5% rotenone).

Antimycin A, an antifungal antibiotic, was first discovered in the United States in 1945 (Leben & Keitt 1948). The first report of biosynthesis, from 2 species of fungus of the genus *Streptomyces*, was probably that of Birch *et al.* (1962), and its potential as a piscicide was first suggested by Derse and Strong (1963). Berger *et al.* (1969) found that a solution of crystalline antimycin in acetone had a good shelf-life, as did a formulation of antimycin on sand grains (1% antimycin, 24% Carbowax and 75% sand). The sand formulation was developed to kill fish by controlled release of antimycin to a depth of 1.5 m or 5 feet (Fintrol 5) and 4.6 m or 15 feet (Fintrol 15). Fintrol bar, a small block of concentrate weighing 250 g and dissolving in about 7 h at 10°C, was developed for use in streams (Lennon & Vezina 1973). Antimycin does not repel fish. Like rotenone, it acts by inhibiting the transfer of electrons along the respiratory pathway (Ahmad *et al.* 1950).

Toxicity to antimycin varies among fish species. Tests with 24 American species at 12°C for 24 h showed that the most sensitive, including rainbow trout (*Salmo gairdneri*), brown trout (*Salmo trutta*) and perch were killed at concentrations of 0.8 μg/litre. Intermediate sensitivity was found in pike, carp (*Cyprinus carpio*), largemouth bass (*Micropterus salmoides*) and pumpkinseed (*Lepomis gibbosus*) which died at 5 μg/litre. The most resistant group included the goldfish (*Carassius auratus*), lethal dose 100 μg/litre, and species of catfish or bullhead (*Ictalurus* spp.), for which the lethal doses varied from 20 to 120 μg/litre depending on the species (Walker *et al.* 1964).

Frogs and tadpoles in field tests with antimycin concentrations of 5–10 μg/litre showed no lethal effects (Schnick 1976). Tiger salamanders (*Ambystoma tigrinum*) survived 80 μg/litre antimycin in the laboratory for 96 h (Walker *et al.* 1964), and Gilderhus *et al.* (1969) found that herons and ducks suffered no ill-effects after exposure to water treated with 10 μg/litre and eating fish killed with antimycin. Rats (*Rattus rattus*) survived a daily dose of 26 mg/kg antimycin (Derse & Strong 1963). The LD_{50} for a dog has been given as 5 mg/kg, for a lamb 1–5 mg/kg and for a rabbit 10 mg/kg (Rieske 1967). A more detailed review of the effects of antimycin on plant and animal life is given in Schnick (1976).

Antimycin in a stream application can be detoxified using one mg/litre potassium permanganate (Ayerst Research Laboratory 1970). Walker (1966) found this concentration toxic to rainbow trout at low temperatures but noted that 0.5 mg/litre successfully deactivated antimycin concentrations of 0.5 to 5.0 μg/litre. Activated charcoal is also an effective deactivator of antimycin (Lennon 1973).

3 Historic use of rotenone as a piscicide

The wide distribution of rotenone-containing plants explains why this toxicant has been used for so long as a piscicide by people of many different cultures. In south-east Asia and Oceania, the poison was known as 'tuba'. Rumph (1747) describes how natives of the East Indies crushed the roots and nodules of *Derris elliptica*, sometimes along with small shrimps, then, mixing it with human dung, placed it in the water to be treated. Marsden (1784) states that natives of Sumatra steeped roots of a creeping plant called 'toobo', of strong narcotic qualities, into water where fish had been seen.

In tropical South America, several genera of Leguminosae are represented, including rotenone-containing *Lonchocarpus* and *Tephrosia* (*Cracca*). The toxicant is known variously as cubé, timbo, barbasco, nicou, canapi, haiari, depending on the plant or the locality. Indians on the Guayaquil River in Ecuador are reported to have chewed the leaves of barbasco, mixed it with bait and scattered it on the river (Juan & de Ulloa 1758). The Creek Indians of the United States used devil's shoestring (*Tephrosia virginiana*) as a fish poison. Flat-topped stakes were hammered into the river at the head of a pool until the tops were level with the water surface. The roots of *Tephrosia* were placed on the stakes and hammered with wooden mallets so that pieces of the crushed plant fell into the water. The affected fish were then killed with spears or arrows (Swanton 1946). Several early accounts of the use of rotenone are mentioned by Krumholz (1948).

4 Present-day application of piscicides

Both rotenone and antimycin are used in the treatment of running and still waters. Treatment of a stream may be undertaken because of a need to eradicate an unwanted species; to reduce recruitment of stream-spawning fish to a lake (Walker 1975); or to limit the spread of disease following the escape of suspected disease-carrying fish from a fish farm. Treatment of a lake may be undertaken to eradicate predators prior to stocking with sport fish; to reduce the size of a fish population; or to eliminate one species but leave the remainder unharmed. Fish population estimates have been made by screening off shallow areas of a lake and treating these with rotenone (Almquist 1959).

With the possible exception of their traditional use by indigenous peoples, the use of piscicides normally requires the authority of national or local government. In the United Kingdom, authority to use piscicides in England and Wales 'for a scientific purpose, or for the purpose of protecting, improving or replacing stocks of fish' is given by the Minister of Agriculture, Fisheries and Food (Salmon and Freshwater Fisheries Act 1975, Part 1, Section 5 (2)). In Scotland, permission has to be obtained from the Secretary of State through the Department of Agriculture and Fisheries for Scotland (Salmon and Freshwater Fisheries (Protection) (Scotland) Act 1951, Part 1, Section 9).

The procedure for applying piscicide to streams is illustrated in Appendix 1. By calculating the volume of water passing a given point during a stated time period, it is possible to work out the volume of piscicide required to treat it at a rate of, for example, 0.5 mg/litre for 30 min. In Method 1, the piscicide is applied in small doses at intervals as the operator walks upstream. In Method 2, 'drip stations' are set up at distances that will enable the required concentration to be maintained.

The procedure for lakes is illustrated in Appendix 2. Ideally, the outflow should be dammed. The volume of the lake is calculated (area × mean depth) and from this, knowing the concentration required, eg 0.05 mg/litre, the total amount of piscicide for the job is estimated. Sufficient piscicide should be obtained to provide additional treatment for reed beds, deep holes and the lower reaches of streams.

The method of treating a given body of water varies in details, but the following procedure is generally applicable.

i. Initial survey. In streams, a check must be made on the accessibility of the areas to be treated and a decision made on how fish downstream from the treatment area are to be protected, eg by the use of potassium permanganate or by relying on the diluting effect of a lake or large river. In lakes, it will be necessary to measure the area and mean depth and note the location of any deep holes or weed beds. A decision has to be made on whether inflow streams are to be treated and how fish in the outflow are to be protected. (If the construction of a dam is suggested, the consequences for downstream users of the water, including the natural fauna and flora, must be taken into account.)
ii. A decision is required on the amount of piscicide to use and the most suitable formulation. The method of application must also be considered and what is to be done with the dead fish. An estimate of the total cost, including time, manpower and the cost of the piscicide, should then be made.
iii. Following the above preliminary steps, it is important to ask whether treatment is practicable. If a decision to go ahead is made, the matter should be discussed with other interested parties (landowners and tenants, fishing proprietors, etc), and a formal application made to the controlling authority.
iv. On the day of the treatment, the person supervising the work should make certain that only the number of people required to do the job are actually involved, and that each person knows exactly what he or she has to do.
v. After treatment has been completed, all equipment used should be removed from the site and empty containers disposed of safely. The site should be visited again during the following few days to make certain that arrangements to protect fish downstream are still working and to record the time when the treated water ceases to be toxic to fish.
vi. The supervisor should write a report about the treatment and include: the reason for it; the type and amount of piscicide used with rate and method of application; details of the areas treated; what was done with the dead fish, noting the species killed and in what proportions, and giving some indication of size groups; what was done with the equipment used; the tests done subsequently to check on the toxicity of the water, and the results of the operation. This record of events should prove useful if there are any enquiries about the work done, and provides a basis for the planning of future treatments.

5 Effects of piscicide treatments on fish and invertebrate populations

The efficiency of a treatment depends largely on maintaining the required concentration of piscicide for the correct period of time and ensuring that it is applied evenly throughout the area to be treated. These points are particularly important when using rotenone, because fish may be able to recover from a sublethal dose. When rotenone was first used by staff at the Freshwater Fisheries Laboratory, Pitlochry, to eradicate pike in Loch Kinardochy, the treatment was unsuccessful, presumably because the concentration used (based on a laboratory assay) was too low at 0.02 mg/litre (Morrison & Struthers 1975). The effects of antimycin are irreversible once the fish shows the first signs of distress (Gilderhus 1972). In lakes, limited treatment of reed beds and insufficient mixing of the toxicant in beds of submerged weeds are reasons for failure to eradicate fish populations. The use of potassium permanganate to protect fish downstream from a treated stretch of river or lake may not be completely effective (Mahon & Balon 1980). Also, potassium permangate detoxifies antimycin rapidly to a level of about 4% of the original concentration and thereafter much more slowly, particularly in water of pH 7.0 (Engstrom-Heg & Colesante 1979).

Only a small proportion of the fish affected by rotenone come to the surface. Bartoo (1977), using marked American yellow perch (*Perca f. flavescens*), estimated that 15.5% surfaced at 24 h and 17.5% at 96 h. It is, therefore, difficult to estimate the biomass of fish in a lake using rotenone. In shallow streams it is less of a problem, although fry may easily be lost in bouldery stretches.

Field treatments with antimycin as a piscicide have confirmed that it can be used selectively. The information available relates to fish species in the United States. No comparable data are available for British fish. Burress and Luhning (1969) showed that antimycin eradicated scale fish (cyprinids, perch, etc) in channel catfish culture, and R. Stinauer (unpublished data) found that 2 µg/litre of antimycin reduced or eliminated gizzard shad (*Dorosoma cepedianum*), carp, bluegill (*Lepomis macrochirus*), long-ear sunfish (*Lepomis megalotis*) and white crappie (*Pomoxis annularis*) without serious harm to largemouth bass in ponds in Illinois. Lennon and Berger (1970) used Finitrol 15 to eradicate heavily parasitized minnows (*Pimephales promelas, P. notatus, Notropis heterolepis, Notemigonus crysoleucas*) in a Minnesota trout lake where thermal stratification had resulted in the minnows being confined to the upper layer. The non-target rainbow trout remained in deeper, cooler water.

It is possible for fish populations to develop a resistance to rotenone treatments. Golden shiners (*Notemigonus crysoleucas*) generally reinfested Ball Pond, New Fairfield, Connecticut, in greater numbers after each of 6 successive rotenone treatments between 1957 and 1974. Paired bioassay tests showed that shiners from Ball Pond were 4–7 times more tolerant of rotenone than shiners from untreated ponds (Orciari 1979).

Concentrations of rotenone high enough to kill in lakes can apparently eliminate certain species of zooplankton from the open water, but populations return to pre-treatment levels within periods of up to 3 years (Brown & Ball 1943; Kiser *et al.* 1963; Anderson 1970; Morrison & Struthers 1975). Applications of 5 µg/litre antimycin to experimental ponds severely reduced numbers of zooplankton (Callaham & Huish 1969). The recovery of zooplankton may be hastened because of reduced predation by fish (Hongve 1977). Many species of zooplankton have stages that are likely to be resistant to the effects of rotenone, eg the so-called resting eggs of *Daphnia* spp., and some copepods spend part of their life history in bottom muds. Anderson (1970) concluded that a zooplankton population that had not reached its reproductive peak was more vulnerable than one that had passed it. One consequence is that a species that recovers rapidly may extend its range temporarily, and occupy the environmental niche normally filled by a species that is recovering more slowly (Kiser *et al.* 1963).

The importance of silt and other organic material to the survival of benthic organisms in waters treated with piscicide was demonstrated by Lindgren (1960). He had noted that tests on the effects of piscicides on invertebrates are often done in the absence of silt or vegetation (eg Hamilton 1941; Chandler 1982). In Lindgren's opinion, 'the rotenone may be neutralized by some kind of adsorption on to mud particles'. This might explain the survival of a range of benthic organisms in Loch Kinardochy, Perthshire, Scotland, which was treated with rotenone on 3 occasions between 1949 and 1972 (Morrison & Struthers 1975). Chandler found that, of the range of invertebrates he tested, all except an ostracod (*Cypridopsis* sp.) were more tolerant of rotenone than most fish. Engstrom-Heg *et al.* (1978) suggest that very few immature aquatic insects in streams could survive 48 h exposure to 3 mg/litre of a 5% rotenone formulation. In Scotland, a concentration of 0.5 mg/litre for 30 min is used in eradicating fish from streams, and observations on the benthos of 3 treated streams showed no significant reduction in benthic populations (Morrison 1977).

Similarly, field studies on the effects of antimycin on benthos have generally indicated that there was little effect at concentrations lethal to fish (Callaham & Huish 1969; Morrison 1979). However, a study of the effects of an application of 10 µg/litre of antimycin to a stream in Arizona showed that numbers, biomass and diversity of invertebrates returned to pre-treatment levels only after a period of 3 years (Minckley & Mihalick 1981).

6 Summary

In managing freshwater fisheries, we may wish to eradicate predators such as pike; reduce the size of a sport fish population to enable the remaining individuals to grow larger; remove fish suspected of carrying disease; or collect samples of fish for research. No one management technique (netting, trapping, etc) is effective in all circumstances, and the eradication of a fish population is usually best achieved with some form of piscicide. Many chemicals which have been used to kill fish are equally toxic to other animals and their use cannot be recommended. Rotenone and antimycin A are highly toxic to fish but less toxic to other vertebrates. Rotenone is the toxic principle in certain leguminous plants, such as *Derris* and *Tephrosia*, and has been used for centuries as a fish poison in Africa and Asia. When used as a piscicide in a lake, it breaks down in water over a period of 2 weeks or more, depending on temperature, after which fish may be introduced. Work done at Pitlochry and elsewhere indicates that rotenone is much less toxic to bottom-living invertebrates than to fish, but zooplankton populations may take several months to recover from treatment. Antimycin A was first isolated in the United States in 1945 from several species of fungi of the genus *Streptomyces*. Its use as a piscicide dates from 1963 and, since then, several formulations have been developed for use in different freshwater habitats. Like rotenone, it is generally less toxic to bottom-living invertebrates than to fish. The rate of breakdown in water depends on temperature and pH, but treated water may be non-toxic within 2 weeks.

7 References

Ahmad, K., Schneider, H. G. & Strong, F. M. 1950. Studies on the biological action of antimycin A. *Archs Biochem.*, **28**, 281–294.

Almquist, E. 1959. Observations on the effect of rotenone emulsives on fish food organisms. *Rep. Inst. Freshwat. Res. Drottningholm*, **40**, 146–160.

Anderson, R. S. 1970. Effects of rotenone on zooplankton communities and a study of their recovery patterns in two mountain lakes in Alberta. *J. Fish. Res. Bd Can.*, **27**, 1335–1356.

Ayerst Research Laboratory. 1970. *Instructions for the use of FINTROL fish toxicant.* New York: Ayerst Research Laboratory.

Bartoo, N. W. 1977. Proportions of a yellow perch population surfacing after a rotenone treatment. *Progve Fish Cult.*, **39**, 27–28.

Berger, B. L., Lennon, R. E. & Hogan, J. W. 1969. *Laboratory studies on antimycin A as a fish toxicant.* (Investigations in fish control no. 26.) Washington, DC: US, USDI Bureau of Sport Fisheries and Wildlife.

Birch, A. J., Cameron, D. W., Harada, Y. & Rickards, R. W. 1962. Studies in relation to biosynthesis. Part XXV Preliminary study of the antimycin A complex. *J. Chem. Soc.*, **1**, 303–305.

Bizyaev, J., Antimov, Y. & Moskalev, Y. 1965. Poisk ihktiotsidov. *Rybovod, Rybolov.*, **8**, 23.

Brown, C. & Ball, R. 1943. An experiment in the use of *Derris* root (rotenone) on the fish and food organisms of Third Sister Lake. *Trans Am. Fish. Soc.*, **72**, 267–284.

Burress, R. M. & Luhning, C. W. 1969. *Field trials of antimycin A as a selected toxicant in channel catfish ponds.* (Investigations in fish control no. 25.) Washington, DC: USDI Bureau of Sport Fisheries and Wildlife.

Callaham, M. A. & Huish, M. T. 1969. Effects of antimycin on plankton and benthic organisms. *Proc. Ann. Conf. South-Eastern Assoc. Game Fish Commiss., 22nd, 1969,* 255–263.

Chandler, J. H. 1982. Toxicity of rotenone to selected aquatic invertebrates and frog larvae. *Progve Fish Cult.*, **44**, 78–80.

Derse, P. H. & Strong, F. M. 1963. Toxicity of antimycin to fish. *Nature, Lond.*, **200**, 600–601.

Engstrom-Heg, R. & Colesante, R. T. 1979. Limitations on potassium permanganate detoxification of antimycin. *N.Y. Fish Game J.*, **26**, 132–141.

Engstrom-Heg, R., Colesante, R. T. & Silco, E. 1978. Rotenone tolerances of stream bottom insects. *N.Y. Fish Game J.*, **25**, 31–41.

Gilderhus, P. A. 1972. Exposure times necessary for antimycin and rotenone to eliminate certain freshwater fish. *J. Fish. Res. Bd Can.*, **29**, 199–202.

Gilderhus, P. A., Berger, B. L. & Lennon. R.E. 1969. *Field trials of antimycin A as a fish toxicant.* (Investigations in fish control no. 27.) Washington, DC: USDI Bureau of Sport Fisheries and Wildlife.

Hamilton, H. 1941. The biological action of rotenone on freshwater animals. *Proc. Iowa Acad. Sci.*, **48**, 467–479.

Hongve, D. 1977. Effects of rotenone treatment on the zooplankton in a small lake. *Vatten*, **33**, 39–42.

Jackson, C. F. 1967. *Detoxification of rotenone-treated water. Pittman-Robertson and Dingell-Johnson Project FW-2-R.* (Technical circular no. 14.) New Hampshire: New Hampshire Fish and Game Department.

Juan, G. & de Ulloa, A. 1758. *A voyage to South America: describing at large the Spanish cities, towns, provinces, etc, on that extensive continent. Interspersed throughout with reflections on the genius, customs, manners, and trade of the inhabitants; together with the natural history of the country. And an account of their gold and silver mines.* London: Davis & Reymers.

Kiser, R. W., Donaldson, J. R. & Olson, P. R. 1963. The effect of rotenone on zooplankton populations in freshwater lakes. *Trans Am. Fish. Soc.*, **92**, 17–24.

Krumholz, L. A. 1948. The use of rotenone in fisheries research. *J. Wildl. Mgmt*, **12**, 305–317.

Lawrence, J. M. 1956. Preliminary results on the use of potassium permanganate to counteract the effects of rotenone on fish. *Progve Fish Cult.*, **18**, 15–21.

Leben, C. & Keitt, G. W. 1948. An antibiotic substance active against certain phytopathogens. *Phytopathology*, **38**, 899–906.

Lennon, R. E. 1973. *The fish control laboratories. Annual report.* La Crosse, Wisconsin: USDI Fish Control Laboratory.

Lennon, R. E. & Berger, B. L. 1970. *A resumé on field applications of antimycin A to control fish.* (Investigations in fish control no. 40.) Washington, DC: USDI Bureau of Sport Fisheries and Wildlife.

Lennon, R. E. & Vezina, C. 1973. Antimycin A, a piscicidal antibiotic. In: *Advances in applied microbiology*, edited by E. Perlman, vol. 16, 55–96. New York: Academic Press.

Lennon, R. E., Hunn, J. B., Schick, R. A. & Burrerr, R. M. 1970. Reclamation of ponds, lakes and streams with fish toxicants: a review. *F.A.O. Fish. tech. Pap.*, **100**, 55–56.

Leonard, J. W. 1939. Notes on the use of derris as a fish poison. *Trans Am. Fish. Soc.*, **58**, 269–280.

Lindgren, P. E. 1960. About the effect of rotenone upon benthonic animals in lakes. *Rep. Inst. Freshwat. Res. Drottningholm*, **41**, 172–183.

Mahon, R. & Balon E. K. 1980. Quantitative collection of fishes from small fast-flowing streams. *Arch. Hydrobiol.*, **90**, 324–329.

Marsden, W. 1784. *The history of Sumatra, containing an account of the government, laws, customs, and manners of the native inhabitants, with a description of the natural productions, and a relation of the ancient political state of that island.* 2nd ed. London: Longman.

Minckley, W. L. & Mihalick, P. 1981. Effects of chemical treatment for fish eradication on stream-dwelling invertebrates. *J. Ariz Nev. Acad. Sci.*, **16**, 79–82.

Morrison, B. R. S. 1977. The effects of rotenone on the invertebrate fauna of three hill streams in Scotland. *Fish Manage.*, **18**, 128–138.

Morrison, B. R. S. 1979. An investigation into the effects of the piscicide antimycin A on the fish and invertebrates of a Scottish stream. *Fish Manage.*, **10**, 111–122.

Morrison, B. R. S. & Struthers, G. 1975. The effects of rotenone on the invertebrate fauna of three Scottish freshwater lochs. *J. Inst. Fish. Manage.*, **6**, 81–91.

Oberg, K. E. 1967. On the principal way of attack of rotenone in fish. *Ark. Zool.*, **18**, 217–220.

Orciari, R. D. 1979. Rotenone resistance of golden shiner (*Notemigonus crysoleucas*) from a periodically reclaimed pond. *Trans Am. Fish. Soc.*, **108**, 641–645.

Pintler, H. E. & Johnson, W. C. 1958. Chemical control of rough fish in the Russian River drainage, California, *Calif. Fish Game*, **44**, 91–124.

Rieske, J. S. 1967. Antimycin A: mechanism of action. In: *Antibiotics*, edited by D. Gottlieb & P. D. Shaw, vol. 1, 542–584. Berlin: Springer Verlag.

Rumph, G. E. 1747. *Herbarium Amboinense.* Vol. 5. Amsterdam.

Schnick, R. A. 1976. *A review of the literature on the use of antimycin in fisheries.* (Rep. no. FWS-LR-74/01.) Springfield, Va: National Technical Information Service, US Department of Commerce.

Stringer, G. E. & McMinn, R. G. 1958. Experiments with toxaphene as a fish poison. *Can. Fish Cult.*, **23**, 39–47.

Swanton, J. R. 1946. The Indians of the Southeastern United States. *Bull. Bur. Am. Ethnol.*, no. 137.

Tang, Y. 1961. The use of saponin to control predaceous fishes in shrimp ponds. *Progve Fish Cult.*, **23**, 43–45.

Walker, A. 1975. The use of rotenone to control recruitment of juvenile brown trout (*Salmo trutta* L.) into an 'overpopulated' loch. *J. Inst. Fish. Manage.*, **6**, 64–72.

Walker, C. R. 1966. (Resource publication no. 17.) Washington, DC: USDI Bureau of Sport Fisheries and Wildlife.

Walker, C. R., Lennon, R. E. & Berger, B. L. 1964. *Preliminary observations on the toxicity of antimycin A to fish and other aquatic animals.* (Investigations in fish control no. 2.) Washington, DC: USDI Bureau of Sport Fisheries and Wildlife.

Appendix 1

Streams may be treated by adding small quantities of piscicide at intervals along their course (Method 1). These then unite to form a long slug of treated water. Alternatively, drip stations may be set up to release the piscicide into the water for the required time (Method 2). In some streams, several drip stations will be necessary to maintain the required concentration. This can be determined by measuring the decrease in concentration of a dye or salt solution released at the upstream site. Fish held in cages at intervals along the stream will verify the effectiveness of the treatment. A potassium permanganate drip may be used to detoxify the piscicide, or if the stream enters a lake or large river the increase in water volume would reduce the piscicide to non-toxic levels.

A characteristic of many substances that are toxic to fish is that a low concentration for a long period of time is often as lethal as a high concentration for a short period of time. Due to the irregular flow of stream water, wild fish populations some distance downstream of the last application point may receive low levels of toxic water for some hours after completion of the treatment, particularly where several treated tributaries combine to form a major stream.

Appendix 2

When treating a lake, the surface area is divided into sectors and each sector is treated with the correct proportion of piscicide. With liquid formulations, a surface application may not be effective at depths greater than 2–3 m. Deeper water may be treated by injecting the piscicide down a weighted length of hosepipe. Reed beds may be treated using a pump with a suitable 'throw'.

Appendix 1

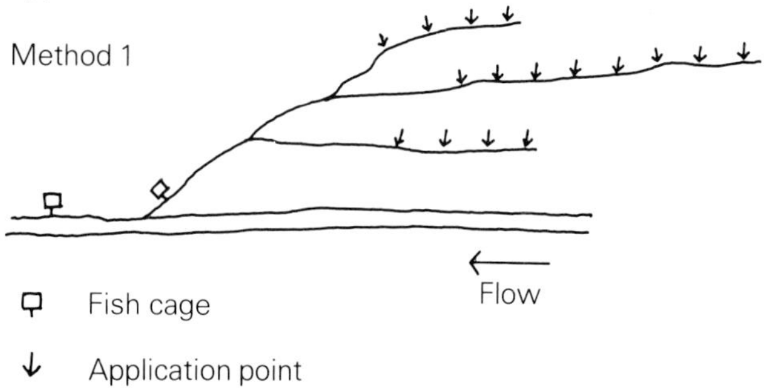

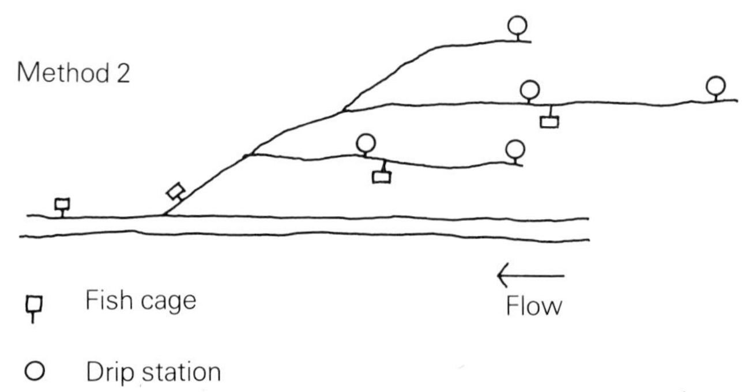

Appendix 2

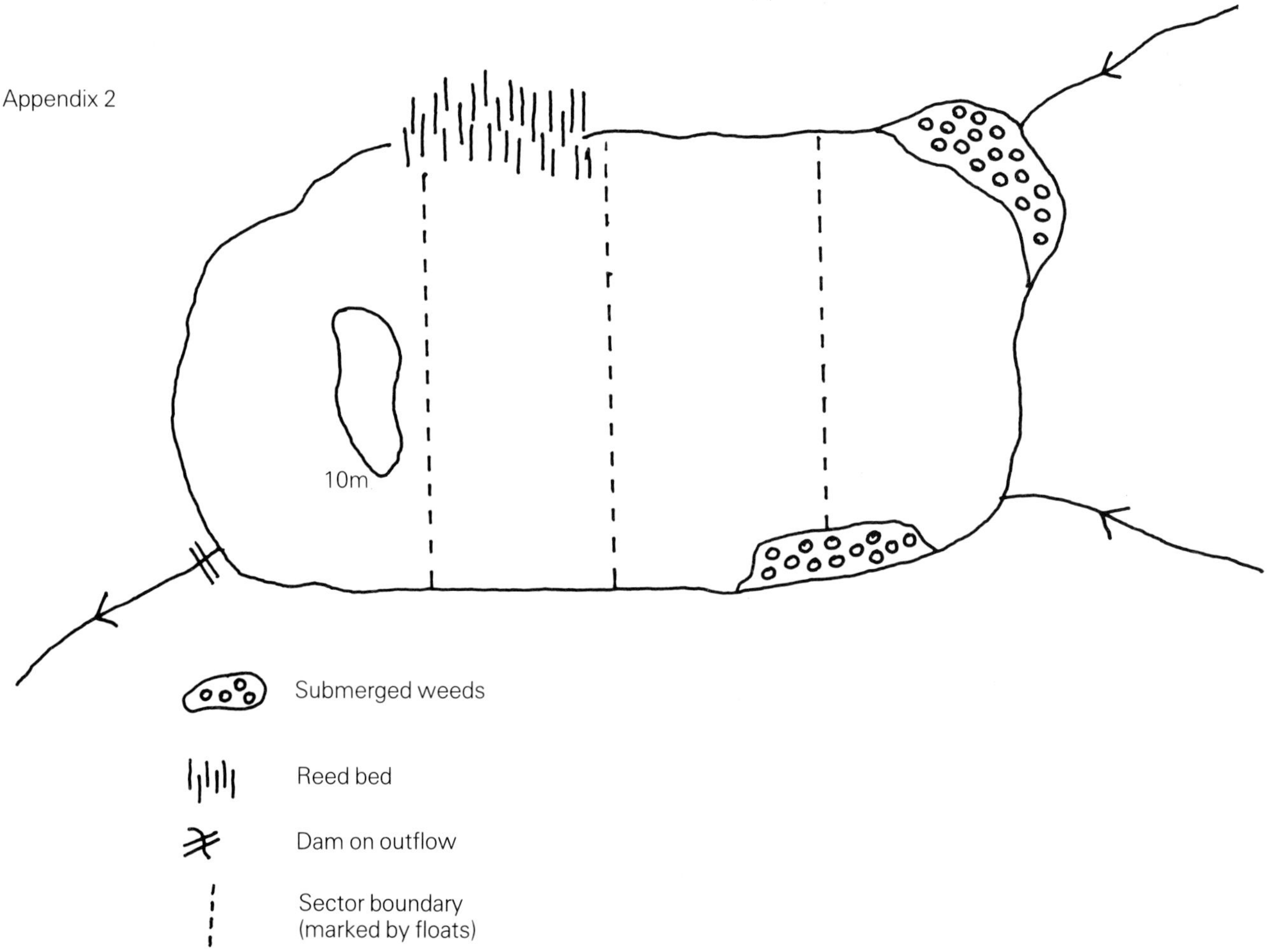

Predator control

D H MILLS
Department of Forestry and Natural Resources, University of Edinburgh, Edinburgh

1 Introduction
Predator control is an emotive subject in which there are few grey areas, and this is particularly true in the realms of fisheries management. One's opinions, however scientifically based, sometimes cause offence. I am well aware that in talking on this subject I am walking on a tightrope, with friends on either side waiting to disown me if I fall into the wrong camp. Feelings tend to run high and this is because there is now a large groundswell of opposition to the culling of animals that are suspected of reducing fish stocks. The time has come when anglers have to review carefully their code of behaviour. They are already criticized by some animal protection societies and are not entirely blameless for the death of many birds ensnared in nylon, choked by hooks, poisoned by lead shot, and scared off their nests by the sound of marching waders. Furthermore, their attitude to the natural environment is not always commendable (Mills 1979).

There was a time when the shooting of raptors and fish-eating birds and the hunting of otters went unquestioned. It was a management practice which was undertaken with satisfaction and a sense of achievement. Anything red in tooth and claw was strung up, as I can well remember witnessing at the numerous gamekeepers' 'larders' which used to festoon the countryside like giant shrikes' 'larders'. We lost many of our rarer bird species in those days, particularly the red kite (*Milvus milvus*) and the osprey (*Pandion haliaetus*), and many of our distinguished 19th century naturalists have much to answer for in this respect. In those days, bird-watching was confined to the clergy, the teaching fraternity and the natural history societies held together by a small but enthusiastic membership. This situation has all changed now. The Royal Society for the Protection of Birds alone has a membership in excess of 300 000 and the Young Ornithologists' Club a membership of over 100 000. There are now also many more wildlife trusts, conservation societies and environmental organizations. Anglers and commercial fishermen are, therefore, very much in the public eye when it comes to predator control, and I would like to consider those areas where problems in this field must be faced – namely rivers, still waters (including reservoirs and small 'put and take' fisheries) and the fish farm, where the angler rears some of his fish prior to their release into angling waters.

2 River fisheries
With more scientific management of fisheries there has grown a large scientific and popular literature on the diet of fish-eating birds and mammals (Mills 1967). With such knowledge and an estimate of the feeding rates (Table 1) of these animals, there developed multiplication sums, involving days of the year and weight or numbers of fish caten daily, which led to horrific estimates of fish consumed annually. Taken a step further, estimates were made to determine what beneficial effect the control of a particular predator would have on fish stocks. The classic work in this field has been carried out in Canada by Elson (1962).

Table 1. Mean daily food intake of some fish-eating animals

Species	Mean daily food intake (g)	Authority
Great crested grebe	340	Harrison and Hollom (1932)
Cormorant	425–700	Macan and Worthington (1951) Mills (1965)
Heron	325	Skokova (1963)
Goosander	400	Mills (1962a)
American merganser	495	White (1957)
Osprey	260	Weir (unpublished)
Grey seal	6800	Steven (1934)

He showed that the control of the American merganser (*Mergus merganser americanus*), when its density on small-to-medium-sized rivers was greater than one bird per 8 ha, could result in a significant increase in the production of Atlantic salmon (*Salmo salar*) smolts. Using the results of Elson's work, I was able to show that control of the closely related goosander (*Mergus merganser*) could similarly benefit some Scottish salmon rivers (Mills 1962a). However, before such a control policy could be widely implemented in eastern Canada, the Canadian Wildlife Service required irrefutable evidence that an increase in salmon smolts resulting from merganser control led to an increased return of adult fish. We all realize how difficult such proof would be to obtain when one considers the sometimes appreciable yearly fluctuations in commercial and angling catches which in themselves do not necessarily indicate the size of the stock returning to the river that season. It is such a reason that those opposing bird control use either when refusing to support a cull of birds or mammals, or when in a position to withold a licence to shoot them. There has been some criticism of Elson's findings and Lack (1966) felt that, while these birds ate many fish, their ultimate influence on the size of the fish population was not, in his view, established. The fishery manager is frequently in a quandary as to what decision to make when, on the one hand, he is faced by this argument and, on the other, he may be told that for every 100 salmon smolts eaten by a goosander 10 were destined to return as adult salmon with an average weight of, say, 4 kg and with a market value of £6/kg. It may be argued that, if the goosanders had not eaten the smolts, some other predator would, or it may be pointed out that these birds also eat other predators of salmon smolts. Larsson and Larsson (1975) considered that in Swedish rivers the loss of hatchery-reared smolts from goosander predation was small when compared with predation by burbot (*Lota lota*), which were estimated to consume a minimum of 175 000 smolts each season. However, goosanders are seen eating smolts; burbot, being under the water, are not.

When I last carried out a survey of its breeding and winter distribution in Scotland in 1961 (Mills 1962b), the goosander tended to breed only in central and northern Scotland but had a much wider winter distribution, with its numbers being supplemented by migrants from Scandinavia. However, since that time, its breeding distribution has extended to southern Scotland and north-west England and into Wales (Meek & Little 1977; Tyler 1985). Current research on the distribution and feeding habits of the goosander and red-breasted merganser (*Mergus serrator)* is being carried out at Durham University for the Nature Conservancy Council, who wishes to ascertain whether or not the breeding range in Britain of these 2 species is still increasing and to obtain more precise information on their densities on Scottish rivers so that a decision can be reached as to where and when to grant licences under the Wildlife and Countryside Act 1981 to shoot these birds.

3 Still waters

Reservoirs and 'put and take' fisheries on small areas of still water have their own problems, usually from cormorants (*Phalacrocorax carbo*) and grey herons (*Ardea cinerea*). There is no doubt that cormorants do eat large numbers of trout (Mills 1965), but shooting is not necessarily the best form of control. There is a danger to anglers and other water users, and usually the shot cormorants are quickly replaced by others. Shooting can also disturb other wildlife. An example of the problems involved is illustrated by an experience at Linlithgow Loch in central Scotland. Here, cormorants visited the water chiefly during the close season when they are not disturbed, and anglers were anxious to shoot them. The loch is also a bird sanctuary. However, it was grudgingly agreed by the Scottish Wildlife Trust and Department of the Environment, who jointly manage the water, to let the anglers shoot a number of cormorants to find out what they had been eating. It was found that they had all been taking perch (*Perca fluviatilis*) just prior to being shot. Perch is a fish that the anglers had been trying to eliminate in this loch because of their competition with the trout for the available food supply. It is, of course, unwise to base one's knowledge of the birds' diet on an analysis of samples taken on one day at one particular time of the year, as there may well be seasonal variations in the diet of any predator depending on what is available. In this instance, more cormorants soon appeared on the loch and there was a significant increase in their numbers shortly after the water was stocked with hatchery-reared trout.

Another bird that can cause problems in still waters is the black-headed gull (*Larus ridibundus*), not so much because of its depredations on fish but because it is responsible for the spread of the eye fluke (*Diplostomum spathaceum*) in trout, particularly rainbow trout (*Salmo gairdneri*). We had this problem in some small man-made ponds near Edinburgh. The gulls nested on an island and on the land surrounding one of the ponds, and they were responsible, together with freshwater snails (*Limnaea* spp.), the primary host, for continuing the life cycle of the fluke which had been introduced to the water by some infected hatchery-reared trout. Shooting seemed pointless and unlikely to be 100 per cent effective, so we stopped them from nesting simply by making it difficult to land. This was done by laying brushwood on the island and some of the shore area and stretching twine over the island and attaching aluminium strips to the twine. It can, therefore, be seen that it is frequently more effective to control the problem birds by looking at their behaviour and seeing how one can best disrupt it to one's advantage.

4 Fish farms

If you want to attract birds to your garden, you put bread out on a table; if you want to increase the variety, you put out monkey nuts, coconuts, fat and raisins. It is not surprising, therefore, that when you build ponds and put in fish you will get fish-eating birds; what is surprising is the astonishment expressed by some fish farmers when this happens. The birds are not to know that they are not meant to come, and an abundant supply of fish may allow them to increase the size of their clutch of eggs and to have more than one brood in a year. If pigeons come to your peas or your flowering broccoli, you cover the plants with nets; similar precautions should be taken by fish farmers.

There are 2 basic approaches to the problem of predator damage at farms, either passively defending the farm by giving it physical protection or attacking the predator. The use of netting is the commonly adopted method in the first category, and shooting in the second. There is no doubt that netting can be very effective, but it does depend on how it is used. Herons can learn to overcome netting protection, if the net is too close to the water surface (Ransom & Beveridge 1983). Wires, too, can deter certain species from alighting on the pond, but they are useless for the wily heron and some feel they may harm the birds. At present, scaring devices include the use of sheets of newspaper, shining metal sheets or strips of aluminium, vibrating tapes, scarecrows and rattles and Buzzbird, the sonic pest control unit which has been most successful in scaring cormorants at the Queen Mother Trout Fishery in Berkshire. Some of these devices are found to be useful, but usually only on a short-term basis.

Again, taking cognisance of animals' feeding habits and then designing the earth ponds in such a way as to prevent them fishing, is a useful practice. For example, one can deepen the sides of the pond and slope the banks steeply so that birds such as herons, which tend to wade into the water and wait for their prey, are unable to reach the fish (Meyer 1981).

A method, not in use in this country but commonly practised in Denmark, is the use of a rotating pump which sprays jets of water on to the pond surface, thus breaking it up and preventing birds from seeing the fish. With many of these methods, the secret is to put them into use as soon as the farm is set up, and not after the birds have got into the habit of coming to the farm. A farm should also be certified as being bird-proof before it is allowed to go into production, as is the case in Denmark.

5 Discussion

Here, then, are at least some ways of solving the problems of predator control, but I'm afraid that, whatever is decided by our lords and masters, it will always remain unsatisfactory to some. However, I think anglers and fishery managers should consider the following statement made by Dr Borgeson at a symposium on *Predator–prey systems in fisheries management* held in Atlanta, Georgia, in 1978.

'Kingfishers, otters, herons, eagles, ospreys, bears, snakes, turtles, mergansers and grackles are some of the non-piscine predators that may impact (sic) populations of desirable game fish or their prey. In previous years these species have been the subject of control efforts but are now valued by our society to the extent that control is largely a thing of the past.'

Also valued by society, of course, are fish – the prey species. In any management of a biological resource, one has the problems of disease, parasitism and predation. It is probably only in the realm of predation that there is any conflict of human feelings, as with orchard owners and bullfinches (*Pyrrhula pyrrhula*), cereal farmers and geese, and commercial fishermen and grey seals (*Halichoerus grypus*). But who is being selfish – the fisherman, the angler, the bird-watcher or the animal lover? I think all should be working to the same end, and the operative word should be 'control'. No-one is suggesting extermination of a species, but if the numbers of one get out of hand to the detriment of the other, in which people have a commercial interest, surely there has to be some control in some form or other – either by culling or deterring? The angler or commercial fisherman should not always be considered the villain of the plot, far from it – he might even be seen to be the fairy godmother. For example, Fisher showed that the dramatic increase in the range and numbers of the fulmar (*Fulmarus glacialus*) was linked to an increase in the number of fishing vessels; herons have increased, probably as a result of an increasing number of fish farms; great crested grebes (*Podiceps cristatus*) have increased as a result of more disused gravel pits being managed as fisheries (Pratt 1977); the fish- and bird-eating mink (*Lutreola lutreola*) have been controlled not by the ornithologist but by the angler, and the future survival of the kingfisher (*Alcedo atthis*) and dipper (*Cinclus cinclus*) has been assured as a result of reduced river pollution – and, as Mellanby (1976) states, it is the angler who is the watchdog on water pollution.

6 Summary

There is now a large groundswell of opposition to the culling of animals that are reported to be affecting fish stocks. The time has come when the angler and fishery manager have to review closely their code of behaviour. They are already strongly criticized by some animal protection societies, and are not entirely blameless for the death of many birds ensnared in nylon, choked by hooks, poisoned by lead shot and scared off their nests by the sound of marching waders. The shooting of many bird and mammal species on the grounds of fish conservation is now being widely questioned. There are a number of areas where problems face the angler, fishery manager and fish farmer in the field of predator control on rivers, reservoirs, 'put and take' fisheries and fish farms. Probably the greatest conflict centres on the control of the increasing number of sawbill ducks which, the anglers say, have a serious effect on juvenile salmon populations, but the bird conservationists contend that this effect has not been proven. In the realms of fish farming, several successful measures have been introduced to deter bird predators, without resorting to killing.

7 References

Borgeson, D. P. 1979. Controlling predator-prey relationships in streams. In: *Predator-prey systems in fisheries management*, edited by H. Clepper, 425–430. Washington DC: Sport Fishing Institute.

Elson, P. F. 1962. Predator-prey relationships between fish-eating birds and Atlantic salmon. *Bull. Fish. Res. Bd Can.*, **133**, 1–87.

Harrison, T. H. & Hollom, P. A. D. 1932. The great-crested grebe enquiry, 1931. *Br. Birds*, **26**, 62–92, 102–131, 142–155, 175–195.

Lack, D. 1966. *Population studies of birds*. Oxford: Oxford University Press.

Larsson, H.-O. & Larsson, P.-O. 1975. Predation pa byusate odlad smolt i Lulealven 1974. *Rep. Swed. Salm. Res. Inst.*, LFTMEDD 9/1975.

Macan, T. T. & Worthington, E. B. 1951. *Life in lakes and rivers*. London: Collins.

Meek, E. P. & Little, B. 1977. The spread of the goosander in Britain and Ireland. *Br. Birds*, **70**, 229–237.

Mellanby, K. 1976. Angling right. *Nature, Lond.*, **260**, 5546–5547.

Meyer, J. 1981. Room for bird and fish. RSPB's survey of heron damage. *Fish Farmer*, **4**, 23–26.

Mills, D. H. 1962a. The goosander and red-breasted merganser as predators of salmon in Scottish waters. *Freshwat. Salm. Fish. Res.*, **29**, 1–10.

Mills, D. H. 1962b. The goosander and red-breasted merganser in Scotland. *A. Rep. Wildfowl Trust*, **13**, 79–92.

Mills, D. H. 1965. The distribution and food of the cormorant in Scottish inland waters. *Freshwat. Salm. Fish. Res.*, **35**, 1–16.

Mills, D. H. 1967. Predation on fish by other animals. In: *The* *biological basis of freshwater fish production*, edited by S. D. Gerking, 377–397. Oxford: Blackwell Scientific.

Mills, D. H. 1979. The impact of angling on the environment (W. J. M. Menzies Memorial Lecture.) *Proc. Inst. Fish. Manage., A Study Course, Nottingham University, 1979*, 2–12.

Pratt, M. 1977. Angling, fishery management and nature conservation. *Fish. Manage.*, **8,** 52–54.

Ransom, K. & Beveridge, M. 1983. Raiders from the skies. *Fish Farmer*, **6**, 22–23.

Skokova, H. H. 1963. On the quantitative study of feeding in piscivorous birds. *Ornitologiya*, **4**, 288–296.

Steven, G. A. 1934. A short investigation into the habits, abundance and species of seals on the north Cornish coast. *J. mar. Biol. Ass. U.K.*, **19**, 489–501.

Tyler, S. 1985. *Sawbills in Wales, 1985*. Sandy, Beds: Royal Society for the Protection of Birds.

White, H. C. 1957. Food and natural history of mergansers on salmon waters in the maritime provinces of Canada. *Bull. Fish. Res. Bd Can.*, **116**, 1–63.

Fish introductions and translocations – their impact in the British Isles

P S MAITLAND
Institute of Terrestrial Ecology, Bush Estate, Penicuik

1 Introduction

The dispersion of fish species across land masses has always been a matter of some debate and controversy. Many people believe in the ability of fish to cross watersheds via inter-catchment connections, migrating waterfowl and other natural means. Others doubt such phenomena and assume that humans are the main factor in moving fish overland. The truth probably lies somewhere between these 2 views, but there seems little doubt that, whatever their involvement in prehistoric times, humans are now a major factor in transporting fish in many countries of the world. The ecological repercussions from many of these introductions have been colossal. The main objective of this paper is to discuss the philosophy of introductions and translocations in the context of experience in the British Isles and elsewhere.

The closing stages of the last ice age (about 10 000 years ago) can be considered as the starting point in any assessment of the distribution of freshwater fish in the British Isles. At its maximum extent, the last major ice cap covered all of the British Isles except parts of southern Ireland and the Midlands and south of England (Figure 1). There were no fish present in any of the areas further north, and indeed it is probable that there were few species in most of the northern unglaciated areas, whose climate must still have been very cold and the glacial rivers unsuitable for aquatic life. The land mass of the British Isles was very different in extent at this time, for Ireland was still connected to Great Britain which, in turn, was still joined to the main continental land mass. As the ice gradually retreated northwards, so did the British Isles assume their present form, the link between Ireland and Great Britain breaking to form the Irish Sea and that between England and the continent to form the English Channel. Before the final separation from the continent, it is likely that all species of purely freshwater fish at present regarded as indigenous to the British Isles had established themselves, at least in the south-east of England. From this area they dispersed themselves at varying rates to other parts of the country.

A broad outline of the fish fauna and the conservation of freshwater fish in the British Isles has been given by Maitland (1974). There, the status and distribution of all species of fish and their value as a resource is reviewed. The value of freshwater fish to the community is far greater than is normally appreciated and includes human consumption, sport, fisheries, amenity, educational and scientific aspects, as well as a valuable store of genetic material. The main pressures on existing fish stocks are related to the effects of fisheries, pollution and land use. The principal trends in the British Isles are away from natural and stable stocks comprising a mixture of native species towards artificially maintained, unstable stocks of a few species of sporting or commercial value. In particular, the rarer, more sensitive, fish species with poor powers of distribution are being eliminated and replaced by commoner, more robust, species with greater powers of dispersal. Many of these sensitive species appear unable to compete with, or cope with predation from, the more successful southern species (Plate 11).

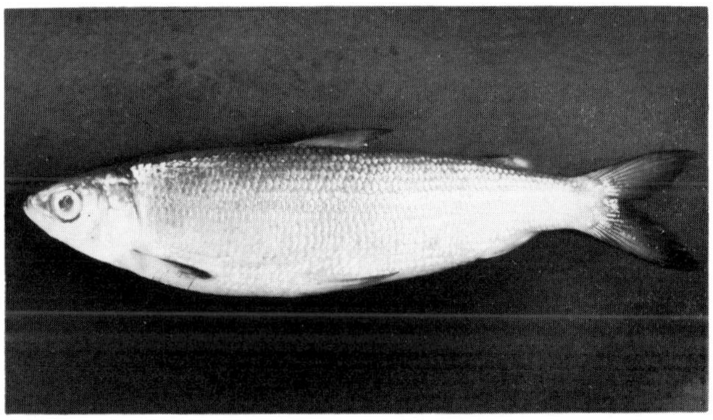

Plate 11. The vendace, one of Britain's rarest species. The introduction of any new species to water containing this fish should be avoided (Photograph P S Maitland)

A second paper (Maitland 1979) considers the rarer species and genetic strains of freshwater fish in the British Isles in some detail. Their present status is reviewed and conservation measures proposed in relation to the priorities involved. Several populations of major importance have already been lost and others are likely to disappear if action is not taken soon. The recent Wildlife and Countryside Act 1981 has drawn attention to the scientific evaluation of conservation sites. In order to produce credible schemes for such assessments, it is necessary to establish national criteria against which the characteristics of any particular site can be evaluated. Maitland (1985) has proposed such criteria for freshwater fish.

2 The British fish fauna

Of the 55 species of freshwater fish found in the British Isles (Table 1), none is endemic, 3 occur only as vagrants, and 12 species have been introduced by humans. Many of the remaining 40 indigenous species are common and widespread, but several are declining in numbers or restricted in distribution. Fish occur in

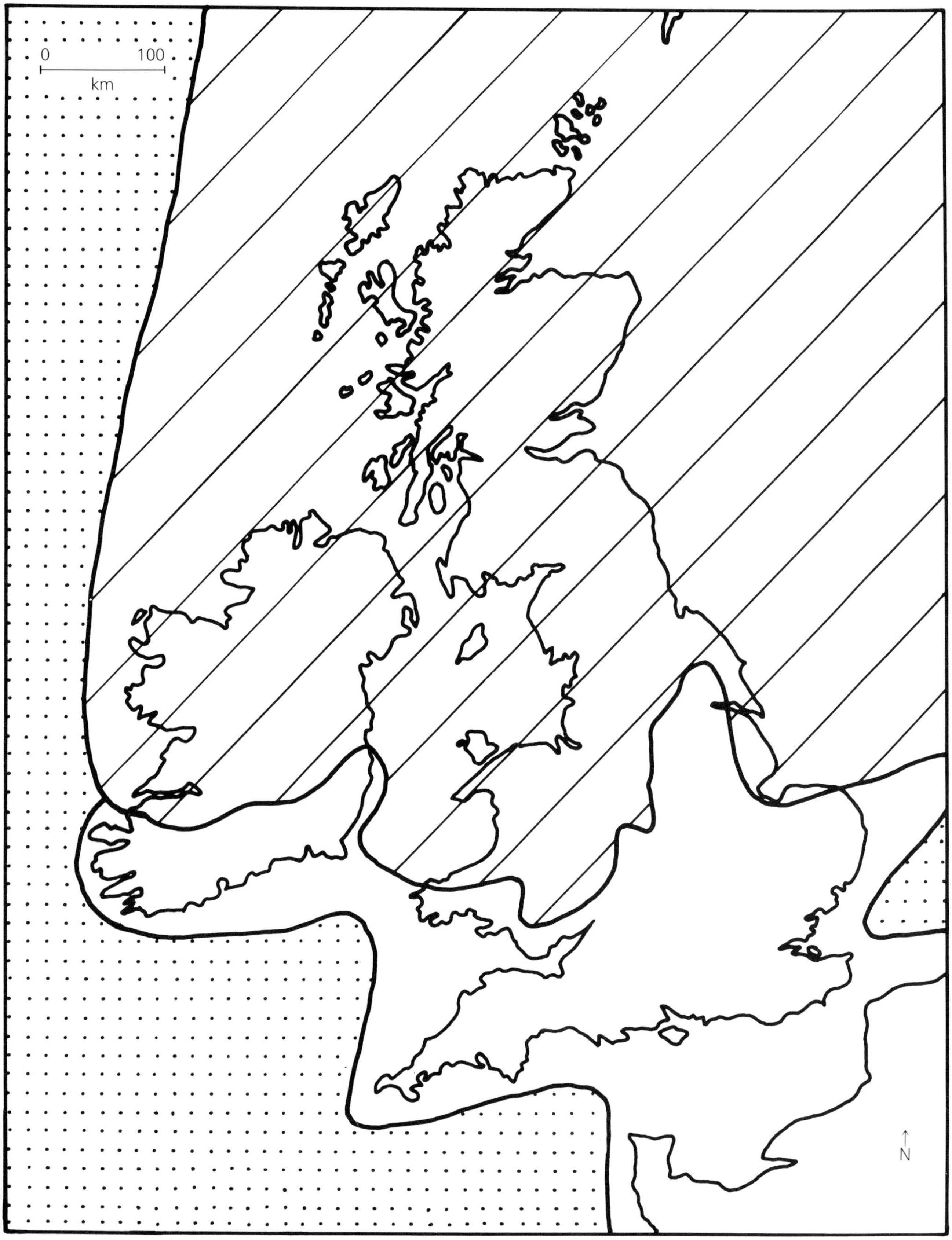

Figure 1. The probable maximum extent of the ice sheet (hatched) covering the British Isles during the last ice age. The approximate boundaries of the sea (stippled) and the land at that time are also indicated

almost all types of open water, except extremely acidic peat pools, grossly polluted waters, high-altitude waters, and those that dry out periodically. Although the general distribution of most freshwater species in the British Isles is now quite well known (Maitland 1969), information is still needed from a number of important sites which it has not yet been possible to examine in detail.

The indigenous fish species of the British Isles may be categorized as follows:

i. those with a marine propensity, ie migratory, including now landlocked migratory species;
ii. those that appear to have mechanisms of dispersal and have moved extensively beyond their original catchments;
iii. those with poor powers of dispersal which are still largely confined to their original catchments.

In general, the number of fish species decreases from south to north in Great Britain: many species are confined to the south and east of the country, and only a few to the north and west. Thus, the maximum number of species which could be expected to occur decreases as one moves north. Some species are well distributed over a large part of the country and found in both running and standing waters, eg salmon (*Salmo salar*), trout (*Salmo trutta* and *Salmo gairdneri*), pike

Table 1. Freshwater fish species established in the British Isles. Closed circles (●) indicate that the species is: A. indigenous, B. of marine origin, C. rare, D. local, E. threatened. Open circles (○) mean the opposite

Common name	Scientific name	A	B	C	D	E
Sea lamprey	*Petromyzon marinus*	●	●	○	○	○
River lamprey	*Lampetra fluviatilis*	●	●	○	○	○
Brook lamprey	*Lampetra planeri*	●	○	○	○	○
Sturgeon	*Acipenser sturio*	●	●	●	●	●
Allis shad	*Alosa alosa*	●	●	●	●	●
Twaite shad	*Alosa fallax*	●	●	●	●	●
Atlantic salmon	*Salmo salar*	●	●	○	○	○
Brown trout	*Salmo trutta*	●	●	○	○	○
Rainbow trout	*Salmo gairdneri*	○	○	○	○	○
Humpback salmon	*Oncorhynchus gorbuscha*	○	○	●	○	○
Arctic charr	*Salvelinus alpinus*	●	●	●	●	●
Brook charr	*Salvelinus fontinalis*	○	○	○	●	○
Whitefish	*Coregonus lavaretus*	●	●	●	●	●
Pollan	*Coregonus autumnalis*	●	●	●	●	●
Vendace	*Coregonus albula*	●	●	●	●	●
Houting	*Coregonus oxyrinchus*	●	●	●	●	●
Grayling	*Thymallus thymallus*	●	○	○	○	○
Smelt	*Osmerus eperlanus*	●	●	●	●	●
Pike	*Esox lucius*	●	○	○	○	○
Carp	*Cyprinus carpio*	○	○	○	○	○
Crucian carp	*Carassius carassius*	○	○	○	○	○
Goldfish	*Carassius auratus*	○	○	○	○	○
Barbel	*Barbus barbus*	●	○	○	○	○
Gudgeon	*Gobio gobio*	●	○	○	○	○
Tench	*Tinca tinca*	●	○	○	○	○
Silver bream	*Blicca bjoerkna*	●	○	○	●	○
Bream	*Abramis brama*	●	○	○	○	○
Bleak	*Alburnus alburnus*	●	○	○	○	○
Minnow	*Phoxinus phoxinus*	●	○	○	○	○
Bitterling	*Rhodeus sericeus*	○	○	○	●	○
Rudd	*Scardinius erythrophthalmus*	●	○	○	○	○
Roach	*Rutilus rutilus*	●	○	○	○	○
Chub	*Leuciscus cephalus*	●	○	○	○	○
Orfe	*Leuciscus idus*	●	○	○	●	○
Dace	*Leuciscus leuciscus*	●	○	○	○	○
Spined loach	*Cobitis taenia*	●	○	○	●	○
Stone loach	*Noemacheilus barbatulus*	●	○	○	○	○
Wels	*Silurus glanis*	○	○	●	●	○
Eel	*Anguilla anguilla*	●	●	○	○	○
Three-spined stickleback	*Gasterosteus aculeatus*	●	●	○	○	○
Ten-spined stickleback	*Pungitius pungitius*	●	●	○	○	○
Burbot	*Lota lota*	●	○	●	●	●
Sea bass	*Dicentrarchus labrax*	●	●	○	○	○
Largemouth bass	*Micropterus salmoides*	○	○	●	●	○
Pumpkinseed	*Lepomis gibbosus*	○	○	●	●	○
Rock bass	*Ambloplites rupestris*	○	○	●	●	○
Perch	*Perca fluviatilis*	●	○	○	○	○
Ruffe	*Gymnocephalus cernua*	●	○	○	○	○
Pikeperch	*Stizostedion lucioperca*	○	○	○	●	○
Common goby	*Pomatoschistus microps*	●	●	○	○	○
Thicklipped mullet	*Crenimugil labrosus*	●	●	○	○	○
Thinlipped mullet	*Chelon ramada*	●	●	○	○	○
Golden mullet	*Chelon auratus*	●	●	○	○	○
Bullhead	*Cottus gobio*	●	●	○	○	○
Flounder	*Platichthys flesus*	●	●	○	○	○

(*Esox lucius*), minnow (*Phoxinus phoxinus*), roach (*Rutilus rutilus*), eel (*Anguilla anguilla*), 3-spined stickleback (*Gasterosteus aculeatus*), 10-spined stickleback (*Pungitius pungitius*) and perch (*Perca fluviatilis*). In a somewhat similar category are those that are common in many sites in the southern half of the country (possibly a few elsewhere), but mostly restricted to standing or very slow-flowing waters (crucian carp (*Carassius carassius),* tench (*Tinca tinca*), bream (*Abramis brama*), silver bream (*Blicca bjoerkna*), rudd (*Scardinius erythrophthalmus*) and chub (*Leuciscus cephalus*)), or to running waters (sea lamprey (*Petromyzon marinus*), river lamprey (*Lampetra fluviatilis*), brook lamprey (*Lampetra planeri*), grayling (*Thymallus thymallus*), gudgeon (*Gobio gobio*), bleak (*Alburnus alburnus*), dace (*Leuciscus leuciscus*), stone loach (*Noemacheilus barbatulus*), ruffe (*Gymnocephalus cernua*), bullhead (*Cottus gobio*) and flounder (*Platichthys flesus*)).

3 Introduced species

An account of the species of freshwater fish introduced to the British Isles has been given by Wheeler and Maitland (1973). The study describes the natural distribution of such fish, the history of their introductions, their success in acclimatization, and their distribution in the British Isles. More than 20 different species are known to have been introduced at various times, but of these only 12 seem to have been successful and have viable populations at present.

The introduction of exotic species of plants and animals to the British Isles was discussed by a working group set up by the Nature Conservancy Council (1979). Unless the reasons are very compelling ones (eg for pest control or food production), the introduction of foreign species is regarded as undesirable, and usually detrimental to nature conservation interests. Thus, it would seem logical to disregard introduced species of fish in nature conservation assessments.

The extensive published information available for Scotland has been reviewed by Maitland (1977), and from this review it is possible to produce some idea of rates of movement of fish into this country (Figure 2). The starting point for Scotland may be taken as the last ice age, say a minimum of 5000 years ago; prior to about 10 000 years ago, the country had been completely covered by ice. The only fish able to colonize as the ice disappeared were those with marine affinities, capable of existing in the ice lakes and glacial rivers which prevailed at the time. Thus, some time later (say 1000 years), there were probably only about 12 species present.

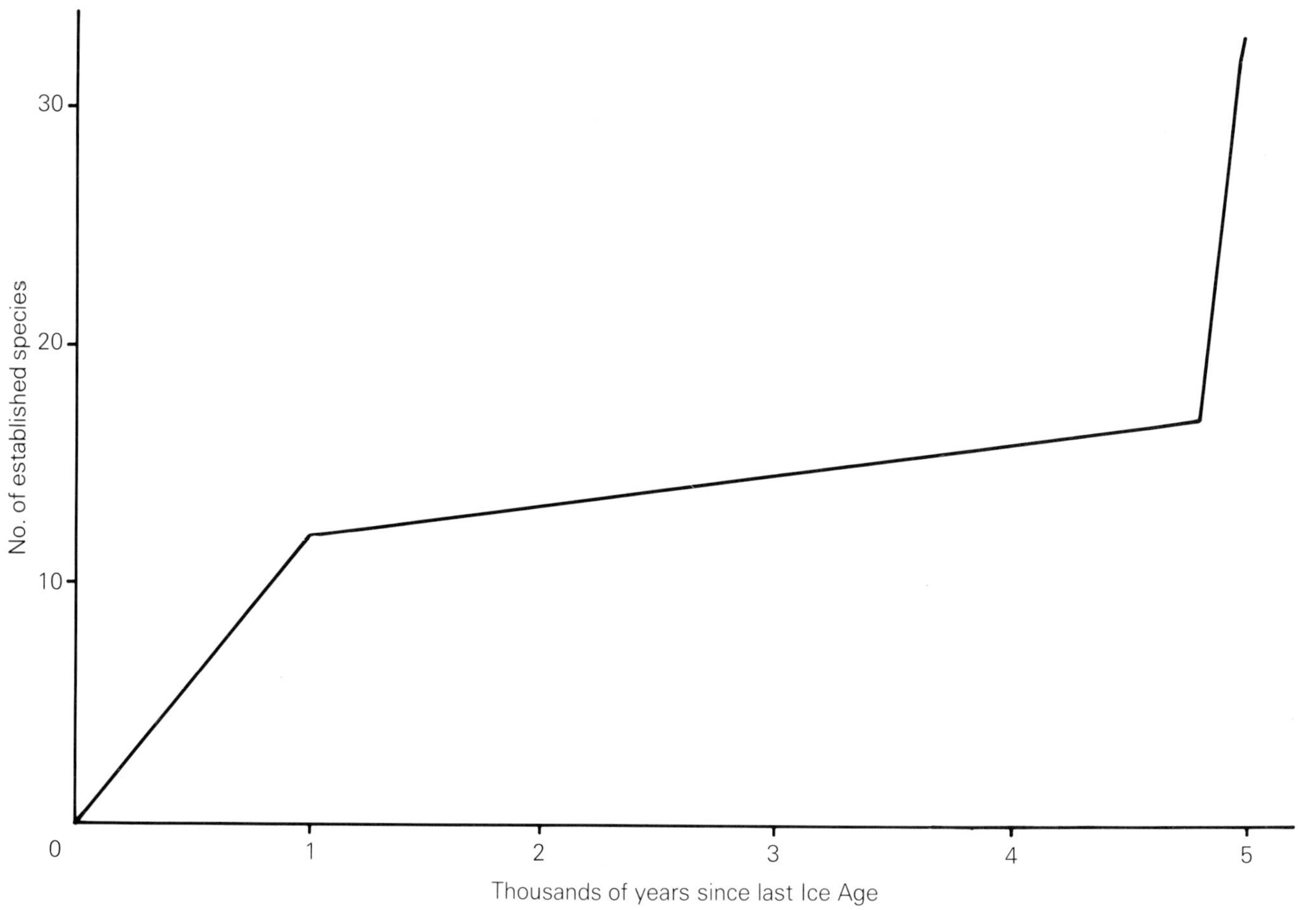

Figure 2. The probable numbers of established species of freshwater fish in Scotland since the disappearance of the last ice sheet, some 5000 years ago. It is assumed that all the present anadromous and catadromous species became established during the first 1000 years

Several thousand years later, by about 1790, only another 5 species had been added to the Scottish fauna: pike, minnow, roach, stone loach and perch. The reasons for the successful movement north of these species are uncertain, but 3 of them (pike, minnow and perch) must have effective powers for dispersal for, apart from species with marine affinities, these are probably the most widely distributed and abundant of British freshwater fish today. The dispersal of fish eggs on the feet of waterfowl is often mentioned as a means of distribution and, while it is true that these 5 species all have adhesive eggs or egg ribbons, there is little real evidence of their dispersal in this or in any other specific way during this period. The role of humans is uncertain, but Campbell (1971) and others believe that they have been responsible for much of the movement of fish in this country.

Some 90 years later (1880), another 5 species were known to occur in Scotland: brook charr (*Salvelinus fontinalis*), grayling, tench, bream (Plate 12) and chub. The main agent of dispersal during this period (and subsequently) was probably human, and there are numerous records of introductions of these and many other species (several of them foreign to the British Isles) around this time. This was a most intensive period of introduction and movement of fish, and many landowners introduced new fish to waters on their estates.

Plate 12. A roach × bream hybrid. One of the many undesirable effects of introducing new species of fish to a water is that they may hybridize with species already present (Photograph P S Maitland)

By 1970, another 8 fish species were known to have established definite populations in Scotland, and for virtually all of them humans seem to have been responsible for the introductions. Many of them are likely to have taken place before 1900; for example, carp (*Cyprinus carpio*) and goldfish (*Carassius auratus*) are mentioned by Scott (1901) as occurring in certain waters in the Glasgow area at that time. Since 1970, another 2 species have been confirmed for Scotland – both probably introduced. Crucian carp actually appear to have been present for some considerable time, whereas ruffe are relatively recent (Maitland *et al.* 1983). It should be noted that many other species were introduced unsuccessfully during these periods.

4 Translocations

For many years there has been an increasing demand from game fishermen, not only for improved fishing for native species such as salmon and trout, but also for introduced species like rainbow trout (*Salmo gairdneri*). A greater interest is also being encouraged in charr, and both the native arctic charr (*Salvelinus alpinus*) and the American brook charr are likely to increase in popularity with anglers. There is also some demand for other North American salmonids which it has not yet been possible to establish in this country.

Especially in Scotland and Ireland (where there is no close season for coarse fish), coarse fishing is increasing rapidly at present due to a combination of increased leisure time, advertizing by tourist agencies and the realization that many of the waters in these countries hold large numbers of specimen fish of various species. This greater interest has brought with it an increasing tendency to move coarse species about (eg Burkel 1971), and more introductions of non-game species are probably taking place now than at any other time this century (Plate 13). The discovery in 1982 (Maitland *et al.* 1983) that ruffe were present and established in Loch Lomond 100 km north of their previous most northerly site was a surprise. Equally surprising has been the apparent enormous increase in their numbers in Loch Lomond since then (Table 2).

Table 2. Numbers of ruffe collected from the trash screens of a water supply system at Loch Lomond

Year	1982	1983	1984	1985
Total numbers	17	47	406	2021
Percentage fish catch	8	19	56	76

5 Stocking

A regular policy of stocking was once considered by many (and is still thought by some) to be the main management tool to be used in fisheries. The large number of derelict fish hatcheries and ponds in many parts of the country provides evidence of the faith of past generations in this procedure. There is no doubt that stocking is a valuable part of the management policy of some fisheries, but it is only necessary where spawning or nursery areas are inadequate to provide recruitment for the angling or commercial fishing pressure involved. The numbers of waters where this is true are in the minority and are likely to remain so, except where 'put and take' fisheries are being developed.

6 Legislation

There have been a number of changes in legislation in recent years, following various attempts to introduce alien fish species to different parts of the United Kingdom. Fortunately, it is now much more difficult to bring foreign temperate fish species into this country and to move them from one part of the country to another. Control is possible through sections included in a number of Acts, the more important of which are listed in Table 3.

Plate 13. The River Endrick near its junction with Loch Lomond. Absent 10 years ago, gudgeon and ruffe introduced here are now among the commonest species (Photograph P S Maitland)

Table 3. The principal pieces of legislation related to the introduction and translocation of freshwater fish in the United Kingdom

Diseases of Fish Act 1937
Salmon and Freshwater Fisheries Act 1975
Import of Live Fish (Scotland) Act 1978
Import of Live Fish (England and Wales) Act 1980
Wildlife and Countryside Act 1981
Diseases of Fish Act 1983

These Acts have a number of strengths and together mean that, as far as the United Kingdom is concerned, it is:

i. illegal to import any live salmon or freshwater fish, or the live eggs of those fishes into Great Britain without a licence under the Diseases of Fish Act 1937;
ii. not an offence to move fish into Scotland from England without a licence, with the exception of Coho salmon (*Oncorhynchus kisutch*, for which an order has been made under the 1978 Act);
iii. unlawful to introduce fish into an inland water in England and Wales (say from Scotland or another Water Authority area) without the written consent of the relevant Water Authority.

Part of the Wildlife and Countryside Act 1981 makes it an offence to introduce any species of fish not at present established in Great Britain to waters in Great Britain. It is also an offence under this Act to introduce the following species: bitterling (*Rhodeus sericeus*), wels (*Silurus glanis*), largemouth bass (*Micropterus salmoides*), pumpkinseed (*Lepomis gibbosus*), rock bass (*Ambloplites rupestris*), and pikeperch (*Stizostedion lucioperca).*

The controls offered by the Diseases of Fish Act might seem to be comprehensive and cover more than the 2 Import of Live Fish Acts do. However, import could only be prevented under the Diseases of Fish Act if the fish to be imported did not meet the Government's current disease criteria; it would be an improper use of the Act to refuse a licence on other grounds. It can be seen, therefore, that the Act does not provide for control of imports *per se*. The Import Acts allow the Secretary of State to make orders in respect of any species whose importation it is considered needs to be prevented or controlled because of environmental or conservation effects.

7 Discussion

It seems likely that there are few aquatic ecosystems into which new species of fish could be introduced and established without some major alteration within the system. Thus, all proposed introductions and translocations of new species, and indeed most stockings of indigenous species, to a water system within the British Isles should be given careful consideration before any action is taken.

Many native species appear to be still gradually distributing themselves throughout the British Isles by natural means (eg pike and perch), but their invasion of new waters may still have dramatic consequences. Several cases are known of pike gaining access to waters containing populations of brown trout and completely eliminating them (Munro 1957). Many species, too, have been dispersed by humans with harmful results. The introduction of roach and dace to the Blackwater system in Ireland, and their influence on other fish there, has been described by Healey (1956), while recently the barbel (*Barbus barbus*) has been moved north in England, well beyond its former area of occurrence. Another controversial translocation in England in recent years has been the pikeperch which has been successfully moved, for angling purposes, outside its previous range. Many hundreds of lesser introductions take place annually, but mostly within the existing distribution boundaries of the species involved.

Undoubtedly, there are some circumstances where the translocation of native species may be justifiable. Obvious cases include introductions to clean waters whose former fish populations have been destroyed by pollution of some kind, and the establishment of suitable species in newly created waters such as reservoirs (Plate 14). Some natural waters contain no fish, and, while it is scientifically desirable that some of these remain without them, a case could be made for the introduction of certain suitable fish into others. It is especially undesirable, on the other hand, that any new species are introduced into waters containing rare native species. Rare native fish may well be moved to other waters as a conservation measure (Maitland 1972).

Where any fish species has become extinct from an area, the original sites from which it has disappeared should always be given first consideration, if reintroduction is contemplated. Extinctions are among the few examples where introductions are fully justified (Nature Conservancy Council 1979), but in a number of cases the conditions at the sites concerned have deteriorated so much that, in the short term at least, alternative sites must be sought to establish new viable populations. Ideally, certain criteria should be followed when selecting such new sites. First, they must offer the full range of habitats necessary for all life stages of the species concerned. Second, the geography of any new sites should be as near as possible to that of the original, and strong preference should be given to those within the same watershed. Finally, the stock chosen for the introduction attempt should normally be from the nearest available site, and certainly from within the British Isles. Precautions should be taken to maximize the genetic diversity of

Plate 14. Loch Awe (left) and Cruachan reservoir (right). One of Britain's rarer fish, the arctic charr, has been accidentally pumped up to the artificial Cruachan reservoir where it now thrives: an example of a useful, if unintentional, introduction (Photograph P S Maitland)

the stock being introduced (Maitland 1979), and introductions from abroad should be considered only as a last resort.

Where fish species new to the British Isles are being considered, several factors should be taken into account before any decision is taken to introduce.

i. There should be a real purpose behind the introduction.
ii. It is normally highly undesirable to introduce species that are likely to have a major influence on the ecosystem.
iii. Consideration should be given as to whether or not the species could be controlled readily.

The rainbow trout is an introduced species which has a very high sporting value and which, though it usually grows faster and may compete with the native brown trout, is never likely to oust it or any other native species because it rarely breeds naturally in this country, and so its numbers can be readily controlled. A similar situation seems likely to hold with the grass carp (*Ctenopharyngodon idella*), which has been introduced to a number of waters (Cross 1969) and is the subject of world-wide interest as a means of controlling aquatic vegetation (Shireman & Smith 1983).

Fish culture, especially of salmonids, has increased enormously in importance in recent years in the British Isles. As well as the evidence of pollution and eutrophication from fish farms which has been detected in some systems, there seems little doubt that caution should also be exercised in relation to damage to wild stocks of fish, particularly that resulting from genetic mixing and change. In developing certain strains of previously wild species of fish to make them more suitable for domestic purposes, they are almost certainly made less suitable for the wild. One of the primary aims of the management of such stocks should be to keep wild and farmed strains separate: this separation is not being done at present. It seems that many of the large farming units for Atlantic salmon in Scotland have many hundreds of thousands of young fish available for sale each year. Many of these fish appear to be sold to private fisheries for stocking in the wild. There are also many escapes to the wild from farm units.

In spite of recent changes in legislation, there is still room for improvement in various areas. The most obvious loophole relates to the aquarium trade, through which it is possible to import, quite legally, a large number of foreign species which could undoubtedly become established in the British Isles. A solution to this problem might be to produce a 'blacklist' of potentially dangerous temperate species which are not allowed into the country without special permits. Another problem, peculiar to Scotland, is the fact that here there is no legislation comparable to that for England and Wales, which aims to prevent fish being moved from one major catchment to another. The introduction of the ruffe to Loch Lomond, described above, is a blatant example of this kind of introduction and parallels, on a lesser scale, the infamous story of the Nile perch (*Lates niloticus*) to Lake Victoria and other African waters (Barel *et al.* 1985). There is no legislation to prevent movement from England into Scotland of species which are held to be indigenous to Great Britain even though they may not be indigenous to Scotland (ie it is not just a matter of lack of control within Scotland).

At the time of writing (April 1986), the Salmon Bill which is being discussed in Parliament has been amended by the Ministry of Agriculture, Fisheries and Food so as to remove control by Water Authorities in England and Wales over introductions into fish farms in their areas. This amendment would have the effect of reducing the extent of Water Authority control over introductions and seems, like several other alterations to or omissions from the Bill, a retrograde step. Welcomme (1986) recently proposed international measures to control introductions.

Finally, in addition to developing adequate legislation, it is extremely important that a social conscience in this area is developed in the public mind – especially among anglers, aquarists, naturalists and others. Only if this education is successful, can we hope to generate the advantages of fish introductions and translocations without receiving many of the potential disadvantages. These codes of conduct among angling and aquarium organizations, in particular, setting out the dangers of moving fish around indiscriminately would be a constructive step towards stabilizing the future of our existing fish fauna.

8 Summary

The nature of fish communities and the fish fauna of fresh waters in the British Isles was originally determined largely by the last glaciation and the separation of southern Britain from the continental land mass. With the retreat of the glaciers, many species with marine affinities invaded most previously glaciated fresh waters, especially barren waters in the north of the country. Thereafter, the dispersion of species was probably a rather slow, chance, process, involving the gradual movement north of many species originally confined to south-eastern England by the glaciers. In recent centuries, humans have had an increasingly important role in the translocation of fish within the country and in the introduction of new species from abroad, and now must be the dominant factor in determining the changing nature of our fish fauna. The fauna is made up at present of 55 established species, 12 of which have been introduced from abroad. At least 12 other species have been introduced unsuccessfully. Many native species have been (and are being) moved successfully into new waters. The latest dramatic example is the ruffe into Loch Lomond, where it may now be the commonest species. Present legislation to control such introductions and translocations, and knowledge of their impact on indigenous stocks are inadequate, and should be reviewed and improved. Equally important is the education of a social conscience in those groups mainly involved in the movement of fish anglers, fishery managers and aquarists and recognized codes of conduct to control the situation should be produced by appropriate governing bodies.

9 Acknowledgements

I am grateful to R N Campbell, R G J Shelton and R B Williamson for constructive comments on the manuscript of this paper.

10 References

Barel, C. D. N., Dorit, R., Greenwood, P. H., Fryer, G., Hughes, N., Jackson, P. B. N., Kawanabe, H., Lowe-McConnell, R. H., Nagoshi, M., Ribbink, A. J., Trewavas, E., Witte, F. & Yamaoka, K. 1985. Destruction of fisheries in Africa's lakes. *Nature, Lond.*, **315**, 19–20.

Burkel, D. L. 1971. Introduction of fish to new water. *Glasg. Nat.*, **18**, 574–575.

Campbell, R. N. 1971. The growth of brown trout *Salmo trutta* L. in northern Scottish lochs with special reference to the improvement of fisheries. *J. Fish. Biol.*, **3**, 1–28.

Cross, D. G. 1969. Aquatic weed control using grass carp. *J. Fish. Biol.*, **1**, 27–30.

Healey, A. 1956. Roach and dace in the Cork Blackwater. *Rep. Sea inld Fish. Eire*, 1956, 3–14.

Maitland, P. S. 1969. A preliminary account of the mapping of the distribution of freshwater fish in the British Isles. *J. Fish. Biol.*, **1**, 45–58.

Maitland, P. S. 1972. *A key to the freshwater fishes of the British Isles.* (Scientific publication no. 27.) Windermere: Freshwater Biological Association.

Maitland, P. S. 1974. The conservation of freshwater fishes in the British Isles. *Biol. Conserv.*, **6**, 7–14.

Maitland, P. S. 1977. Freshwater fish in Scotland in the 18th, 19th and 20th Centuries. *Biol. Conserv.*, **12**, 265–278.

Maitland, P. S. 1979. The status and conservation of rare freshwater fishes in the British Isles. *Proc. Freshwat. Fish. Conf.*, **1**, 237–248.

Maitland, P. S. 1985. Criteria for the selection of important sites for freshwater fish in the British Isles. *Biol. Conserv.*, **31**, 335–353.

Maitland, P. S., East, K. & Morris, K. H. 1983. Ruffe, *Gymnocephalus cernua* (L.), new to Scotland, in Loch Lomond. *Scott. Nat.*, 7–9.

Munro, W. R. 1957. The pike of Loch Choin. *Freshwat. Salmn Fish. Res., Edinb.*, **16**, 1–16.

Nature Conservancy Council. 1979. *Wildlife introductions to Great Britain.* London: NCC.

Scott, T. 1901. The marine and freshwater fishes. In: *Fauna, flora and geology of the Clyde area*, edited by G. F. S. Elliott, M. Laurie & J. B. Murdoch. Glasgow: University Press.

Shireman, J. V. & Smith, C. R. 1983. Synopsis of biological data on the grass carp, *Ctenopharyngodon idella* (Cuvier and Valenciennes 1844). *FAO Fish. Synopsis*, **135**, 1–86.

Welcomme, R. L. 1986. International measures for the control of introductions of aquatic organisms. *Fisheries*, **11**, 4–9.

Wheeler, A. & Maitland, P. S. 1973. The scarcer freshwater fishes of the British Isles. 1. Introduced species. *J. Fish. Biol.*, **5**, 49–68.

Multi-purpose use of waters

M L PARRY
Severn Trent Water Authority, Birmingham

1 Introduction
Alistair Cooke has reminded us that James Madison was once accused of seeking to make human frailty the basis of good government. He replied, 'I know of no other'. The essence of multi-purpose use of water space may almost rest upon human perfection. Any gap between aspiration and achievement is referable to human frailty, and this becomes a factor in all our calculations.

I work for a Water Authority, and Water Authorities are much concerned with multi-purpose use. Most of what I have to say refers to that situation. We are not alone in taking the view that the conservation of wildlife and recreational use can co-exist. 'Management for nature conservation can be compatible with access by the public to enjoy the special character of an area and for a range of recreational uses provided that these activities are suitably controlled to avoid damage to the nature conservation interests' (Ratcliffe 1977).

The meeting at which this paper was originally given primarily discussed angling, and anglers have always accepted the need for discipline and control in carrying out their sport.

A word, then, about the genesis of multiple use – I mean multi-use, not over-use – of Water Authority reservoirs in England and Wales. The Water Act 1973 which created the reservoirs laid upon them a duty (Section 20) to make the best use for purposes of recreation of their waters and lands associated with waters. The Wildlife and Countryside Act 1981, by modifying other sections of the Water Act, requires Water Authorities to further the conservation of wildlife *inter alia* in formulating and carrying out any of their statutory functions, of which the provision of recreation is one. Allowing for human frailty, resolving these 2 inherently conflicting requirements becomes in itself an objective. I want to speak of the policies, the methods and the results, the last of these by reference to a few case histories.

2 Policy
Water Authorities were reconstituted in 1983 and among the changes was a break with the local government-based structure of their previous 10 years. The deemed loss of public contact was made up by establishing a consultative network of committees, one of which is the Recreation and Conservation Committee. In Severn Trent, this committee consists of 22 members to serve 8000 square miles, and there was much agonizing at its conception over the question of the balance of conservation as against active recreation. The solution has been to depart from the majority vote system and to report on the views of the various interests. So far, it appears to have worked very well indeed. The remit of the Committee is to advise the Board, but usually executive action follows from their recommendations without taking the matter to Board level. Common ground is far more common than the pessimists had supposed, and I suspect that Madison had something similar in mind 200 years ago.

It is fair to say that the great bulk of development had already taken place by 1983, along with the policy guidelines for development of recreation that have served for 9½ years. There are, however, still a number of management and occasionally capital development decisions to be taken. Any scheme for using any land or water space for recreation must satisfy all of the following criteria:

i. suitability of the site;
ii. demand for the specific type of facility in the area;
iii. benefit to cost ratio, particularly in relation to the number of users;
iv. compatibility with operational uses and conservation considerations (the conservation requirement is new);
v. likely benefit to the Authority image (and why not?).

Clearly, this filtering system can also be applied by allocating points to produce a ranking of merit for various proposals, if there happens to be competition for finance.

There is also a general policy to lease to user associations or commercial interests, *where appropriate*, the recreational facilities on the Authority's lands and reservoirs so that the management is passed to another organization. The main problem arising from this arrangement is, at least, that any protection for an interest other than the one actually being leased out must be incorporated by way of clauses in the lease, although, of course, one seeks to achieve most of the fine tuning by common agreement. It is, however, a fact of life that a lessee who may have invested considerable funds requires to know exactly where he stands, and the lease document is the wall to get his back against. However, it is a cumbersome and not very flexible way of coping with a situation which is, in essence, flexible.

Objectives change with time, and new information about wildlife, or new priorities for providing either recreation or conservation, are always coming forward; it is extremely difficult to respond to such changes when much of the public use of the water surface is governed by a rigid legal document. Although, therefore, leasing out of a facilities is good news from the point of view of minimizing the Authority's involvement in unwanted local site management, it is not necessarily good news for adjusting the delicate interplay between competing requirements of multi-purpose use. It introduces rigidity where flexibility is required.

More recently, the Authority has also agreed landscape and conservation policy guidelines to complement the above. They depend upon:

i. an evaluation of the site for conservation;
ii. the choices of action to be considered;
iii. compatibility with the main objectives of the land-holding and with recreation policy;
iv. apportionment of the facility between users with perhaps a prime user element;
v. consistency with a sound overall conservation policy.

It follows, then, that when any new facility is being considered there is an iterative process which leads to a final apportionment for such multi-purpose use as its size and nature will permit.

3 Methods

I will not deal with the family of conservation-oriented activity that arises from physical engineering works like river drainage. The relationship between active recreation and the conservation of wildlife on a site is influenced by the timing, location and obtrusiveness of the activity(ies) and the space/time requirements of the animals or plants to be conserved – and their relative vulnerability. It is sensible to regard all types of recreation and the conservation of wildlife together as forms of land use, and to turn aside for a moment from a rigid classification into one or the other. There is really a continuous spectrum, and the more tranquillity-dependent these users are, the more easily they are displaced. For instance, roosting or feeding wildfowl and the bird-watchers who enjoy their presence need protection from disturbance from anglers who, in turn, need protection from sailors who may, in turn, need protection from motor-boaters or water-skiers. Manning (1980) recalls that in 4 rivers in Vermont 'fishermen differ from other user groups in being more traditional, older, relatively declining in numbers, and experience more management problems and conflicts, express lower levels of satisfaction and lower tolerance of higher use tendencies'.

Time zoning can be very effective between sports, and sometimes between use for recreation and for conservation. Space zoning is more popular in resolving this type of potential impact. The concept of a refuge area for birds is very well proven. Watmough (1982) attaches great importance to this idea, as can be seen by his observations on 2 Derbyshire reservoirs which are used for fishing, sailing and walking, as well as being important for wildfowl conservation. Any method adopted is inevitably vulnerable to human frailty and, in this event, simply to those people who will not obey the rules and insist on trespassing where they are not supposed to go. These rules need to be generally adopted by common consent and persuasion, but they work better if there is a real sanction. For this reason, users with something to lose are the more welcome ones. A fisherman with a season permit is less likely (one might think) to hazard his privilege by trespassing than someone who is a mere day permit-holder and he, in his turn, is a better bet than the casual walker or other visitor with nothing to lose at all. Clubs who can be relied upon to enforce their own rules are, therefore, quite popular tenants.

It is a matter of some regret that more is not done to monitor the actual success of these arrangements, not only because all observations involve much effort, but also because interpretation of these always presents problems. Birds ought to be easy to monitor as compared with fish: they are, after all, easy to see and their reactions to disturbance equally so. However, the interpretation of the long-term significance of such observations does offer substantial difficulties.

4 Examples

Figure 1 shows the number of activities that take place on Severn Trent reservoirs. The most frequent is nature study (24), followed by game fishing (21), informal walking, picknicking (14), sailing (9), coarse fishing (7), sub aqua (5), canoeing (3). There are also sundry minor activities.

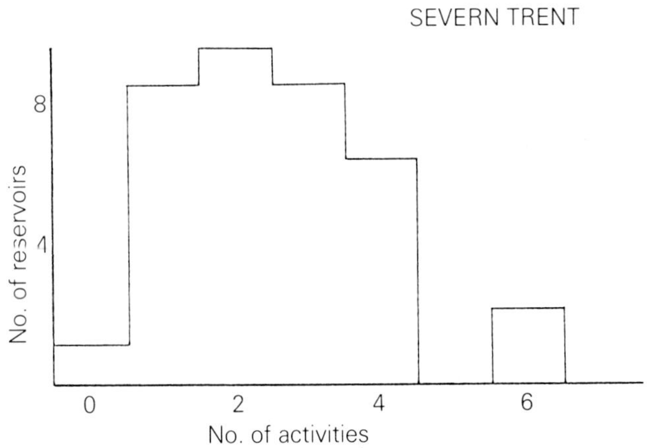

Figure 1. The number of different activities on 34 Severn Trent reservoirs

Birds need sites for nesting, feeding young, feeding as adults, moulting, loafing and roosting. These activities take place in different locations at different times of the year. Birds are also more vulnerable to disturbance than other forms of wildlife associated with reservoirs. It is almost certain that any activity by or on the water will have some impact somewhere, and this impact is rarely one of habitat degradation (eg the lead and swans situation). It is more usually one of disturbance.

It is very hard to decide on the importance of the latter in some cases, but Tydeman (1977) recorded the importance of the coarse fish close season on breeding bird numbers at certain gravel pits, by comparing the situation in 2 successive seasons without and with a close season. Not merely waterside birds but also passerine birds were affected: 4 species of waterside bird increased and one decreased, while 20 passerine species increased and 2 decreased. There were several records of birds breeding there for the first time. There was also evidence that damage to vegetation during the growing season was, in this case, a factor. We can also observe waterfowl roosting or feeding on water areas taking flight when disturbed, and counts of numbers of birds on reservoirs can be very revealing. Owen (1983) mentions the great importance of man-made wetlands in providing roosts and feeding areas, and records the steady rise in the count of wildfowl using the English and Welsh waters in winter over the past 20 years (this record refers to important species of inland wintering wildfowl). Nine species have increased, 3 have remained steady, and the only decline has been locally in the mute swan (*Cygnus olor*). Assessment of the long-term impact of any disturbance pattern on the migrating species is very hard to measure. Those species with a home range are much easier to study in this respect. I must admit that I always encounter great difficulty in getting professional conservationists to talk in terms of the number of these species that ought to be a target for conservation. The implication, obviously useful in management terms, would be that once these numbers are reached there is no longer any continuing need for other types of activity to defer to them, and no need to keep areas quiet. It might also be thought that, in the absence of some such targets, certain species might become so numerous as to risk succumbing to disease. As conservationists will not tell me what population level is needed, I have to try a different approach.

Ladybower reservoir in Derbyshire provides a case in point. This is a large reservoir of some 160 ha. The whole valley is a breeding site for the common sandpiper (*Actitis hypoleucos*) and, when the valley was part inundated by the construction of dams, the birds simply began to breed round the shingle banks of the lake shores.

Holland *et al.* (1982) recorded breeding behaviour and the feeding sites of these birds. The lake is used by boat and bank game fishermen and the mechanism of interference is therefore variable with the season. Breeding and feeding sites are shown in Figure 2, along with the location of anglers. Access to stony areas of foreshore is vital to feeding the young of these birds. Their survival depends upon their weight increasing from 8 g to 20 g in the space of 3 weeks after hatching. The presence of shore anglers in feeding areas near to nests is very important during this phase. It is thought unlikely that they would affect the adults during their mating, laying or incubating phase, because the nests are some way back from the shore.

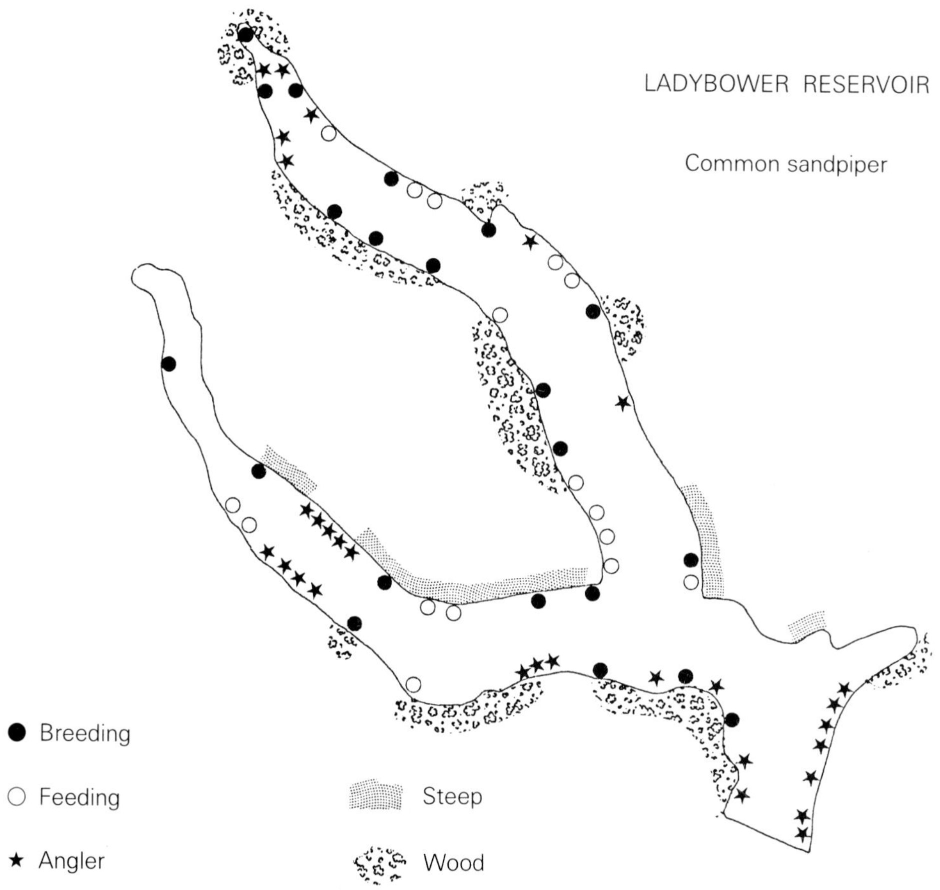

Figure 2. Feeding and breeding sites of the common sandpiper on Ladybower reservoir and sites used by anglers

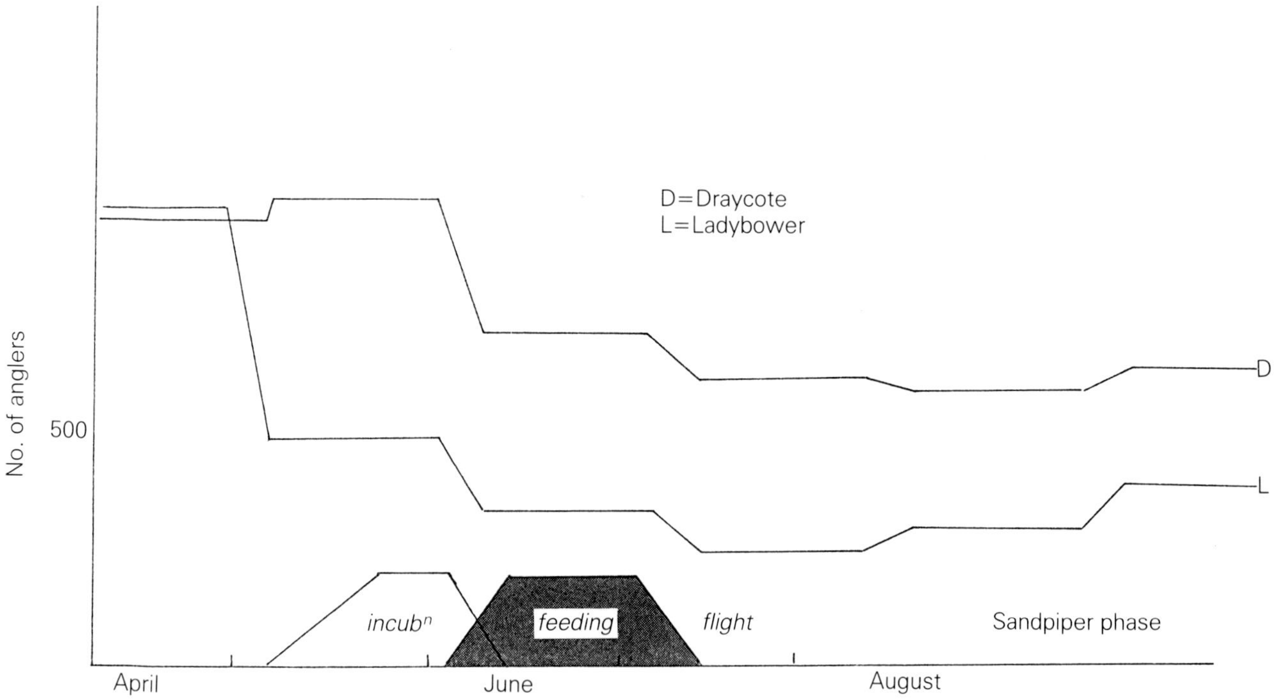

Figure 3. The number of anglers at Draycote and Ladybower reservoirs in relation to the sandpiper breeding season

After flight, the juveniles are not vulnerable and soon leave the area anyway.

Figure 3 shows the numbers of anglers attending Ladybower reservoir during the game fishing months. Numbers are considerably reduced in June when juvenile feeding of the sandpiper is most important. It may, therefore, be possible to set aside areas of foreshore for the benefit of the birds during this month when they most require it and when anglers require it somewhat less.

As a working principle, therefore, it may be that we can aim to achieve not the ultimate for the anglers and not the ultimate for the sandpipers, but a good viable facility for both and, important in human terms, the presence of sandpipers is certainly an interesting addition to the visitor experience. This principle is an exceedingly important one and is often referred to, by me at any rate, as 'viability'. It is important that, in seeking to achieve multi-use, the facility must be large enough and of a character that permits all the proposed activities to take place to an effective extent. If the fishing is impaired so that attendance increasingly drops off, or if the numbers of sandpipers go into a successive decline, then this requirement has not been met — always provided, of course, that other factors are not called into play.

Taking another large waterbody (Draycote, 240 ha, in lowland Warwickshire) as an example: although in the original planning a fairly high degree of recreational activity was planned for this reservoir, leaving the other reservoir of the former Rugby Joint Water Board, Stanford, as a bird refuge, the birds nonetheless persist in using Draycote in considerable numbers. Figure 4 is based on figures provided by G Harrison

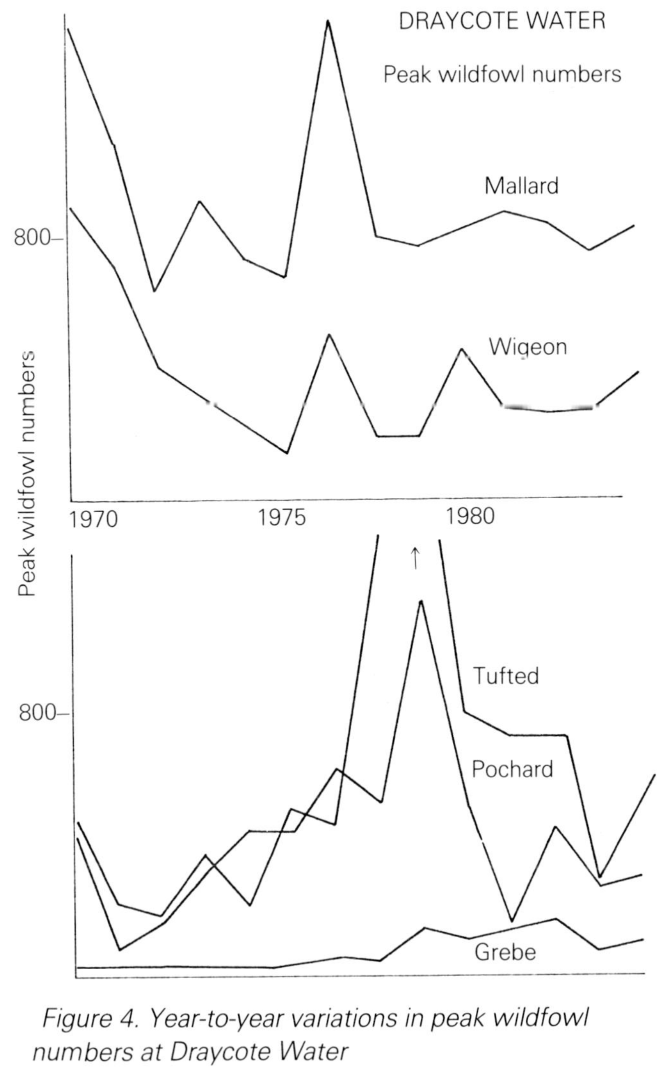

Figure 4. Year-to-year variations in peak wildfowl numbers at Draycote Water

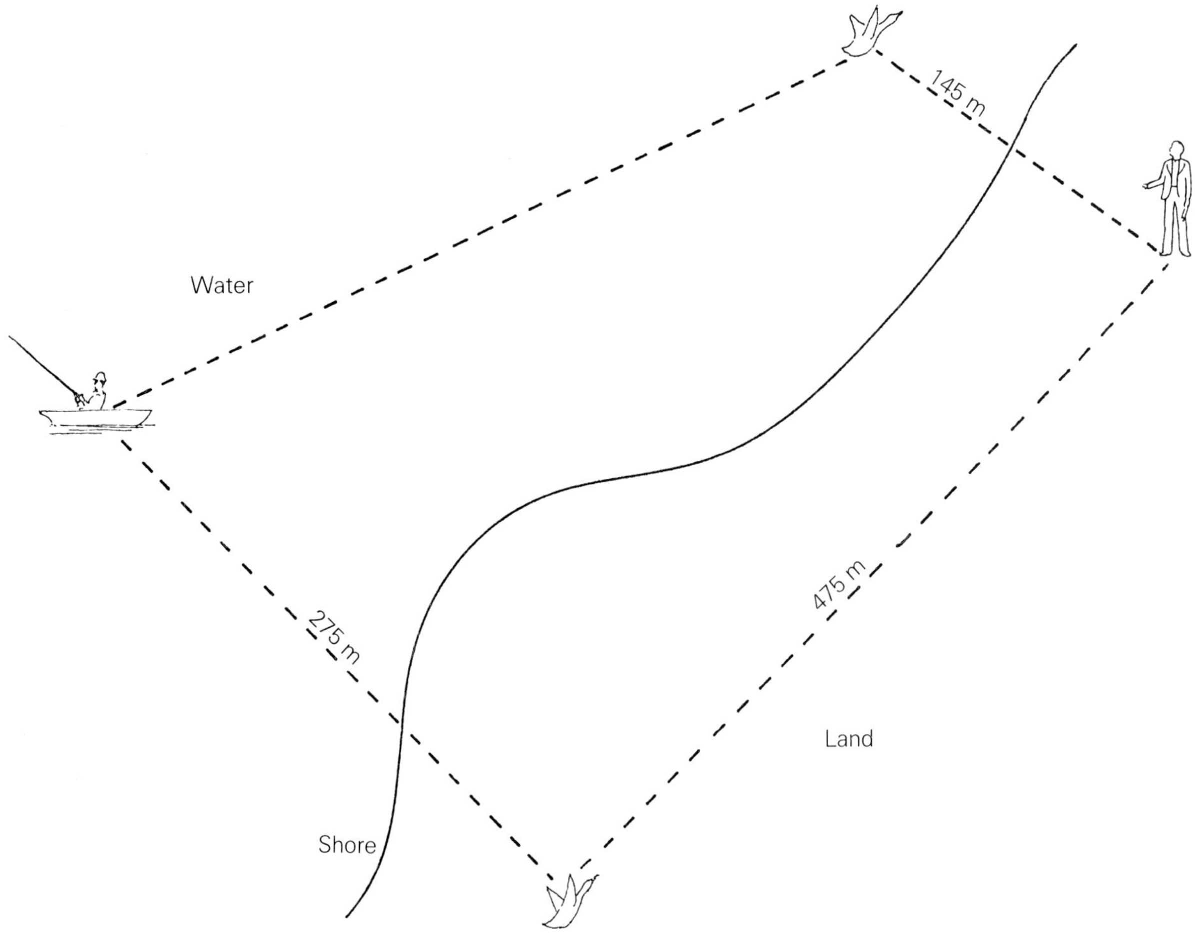

Figure 5. Distances from disturbances at which mallard take flight

and shows how the numbers of 6 species have progressed from year to year. To some extent, the fluctuations are referrable to the natural development of a new inundated water area, but walking and bird-watching around the perimeter have increased since 1978, and we are having to look carefully to see whether these factors are contributing to the decline in numbers over the past 2 or 3 seasons. Birds having alternative places to go, for instance wigeon (*Anas penelope*), change grazing areas if disturbed from the shoreline, and therefore their peak numbers show no decline, although the peaks nonetheless occur after the fishing season which ends on 25 October. It is unfortunately not feasible, due to the proximity of the perimeter track at Draycote, to have a fully effective refuge area.

Such an area, however, is extremely important. At Staunton Harold reservoir in Derbyshire, there is a long arm which is very effective as a refuge zone. The other main activities on the water are sailing and coarse fishing, but both are confined to the wider part of the reservoir.

In planning such refuge areas, the scare distance (Figure 5) is an important consideration (Watmough 1982). The scare distances themselves, however, vary with the season. In hard weather, birds are much more tolerant, but by the same token, if they are caused to fly, the drain on their energy budget can become serious. The loss of feeding time due to disturbance is probably even more important in winter conditions.

The best arrangement from the point of view of conserving wildlife is to have weekend recreation and a good refuge area, but the latter is still effective even if recreation is almost continuous. The impact on numbers of birds seen on the water, however, is not the same as an index of the overall impact of recreation on the total population.

The River Avon in Warwickshire is used extensively for motor-cruising. The effects of boating and river engineering on macrophytes and associated species cause harm, but one cannot demonstrate that fishing is causing any damage to the wildlife interest. Trampling by anglers along the riverside may affect the vegetation, but may also have the effect of increasing the species diversity by eliminating the shady cover afforded by dominant species like nettles (*Urtica dioica*) and great willowherb (*Epilobium hirsutum*) because the pressure is fairly light and intermittent. Fishing pitches, however, are often quite seriously eroded but the effect is local. There is an impression among the users of cruising boats that the river is 'teeming with wildlife', but this is true in fact only of species that are fairly tolerant of humans. The shier species have tended to disappear since the river was opened in 1974 between Stratford and Evesham to motor-cruising. This development demonstrates, again, the 'order of vulnerability' argument.

5 Conclusions

There is a good deal of information on the vulnerability and the robustness of plants and invertebrates, enough to encourage viable multi-use of many waters to be planned fairly intelligently. It is necessary that such planning is flexible, and it is therefore important that management plans are no more rigid than is necessary. It is important for the vulnerable end of the user spectrum, whether of recreational activities or wildlife, that these planning rules are, in fact, carefully drawn up and followed, and an appropriate management structure is essential. Too harsh a regimen in this respect can, if carried to excess, destroy what it seeks to preserve in terms of human enjoyment. It is very doubtful whether this planning should therefore be carried to the extent of being enshrined in actual byelaws or other apparatus involving the panoply of the law. The current proposal to consider a byelaw banning the use of lead for angling must be seen in this context.

6 Summary

The advantages and the problems of multi-use of waters have been experienced in several permutations by Water Authorities, because these bodies have to satisfy statutory requirements to use their waters both for recreation and also to further the conservation of wildlife in all that they do. The policy which the Severn Trent Water Authority has developed in response to these 2 requirements is described. Experiences on some of the waters managed by STWA are presented, and the results of research on the action of recreation on wildlife are mentioned. It is necessary to choose activities and conservation objectives with care. The concepts of 'compensatory conservation' which uses the potential of a site to reinforce some conservation aspect to compensate for the deterioration of another, and of viability so that any activity or conservation objective selected should be of worthwhile proportions, are introduced and discussed.

7 References

Holland, P. K., Robson, J. E. & Yalden, D. W. 1982. The status and distribution of the common sandpiper in the Peak District. *Naturalist*, **107**, 77–86.

Manning, R. E. 1980. Behavioural characteristics of fishermen and other recreationalists on four Vermont rivers. *Proc. FAO conference, Vichy.* Rome: Food and Agricultural Organisation.

Owen, M. 1983. The importance of man-made wetlands to wildfowl wintering in Britain. *Proc. Symp. Wildlife on Man-made Wetlands*, 73–79. Milton Keynes: ARC Wildfowl Centre.

Ratcliffe, D. A. 1977. *A nature conservation review.* Cambridge: Cambridge University Press.

Tydeman, C. 1977. Importance of close fishing season to breeding bird communities on three gravel pits. *J. environ. Manage.*, **5**, 289–296.

Watmough, B. 1982. *Report to Severn Trent Water Authority and Water Space Amenity Commission.* Birmingham: STWA.

Angling and wildlife conservation

D W MACKAY
Clyde River Purification Board, East Kilbride

1 Introduction

Angling and wildlife conservation increasingly appear to be inimical to each other. There are undoubtedly genuine conflicts of interest which are not resolvable by consultation and better understanding, but it does seem that there is scope for more co-operation and less antagonism between the groups concerned with the fresh waters of Great Britain. Interests range from scuba-diving in reservoirs, through canoeing in torrents, to angling in all its forms, and the conservation and study of animals and plants. Unfortunately, in British society there is a tendency towards the adoption of 'purer than thou' attitudes, leading to snobbish behaviour, a degree of arrogance, and, inevitably, conflict. Within angling circles, the dry-fly fishermen often consider their art to be 'purer' than that of the wet-fly or bait fishermen. The salmon angler feels superior to the trout fisherman and coarse fisher in that order, and so on. Outside angling, those whose interest lies in the conservation of rare species look down on those who conserve, for example, game birds and wildfowl with a view to harvesting a proportion of them. Anglers and bird-watchers alike abhor those whose interests result in noise and movement, the ultimate example being the water-skier behind a high-powered boat. This paper attempts to defend the angler within the context of conservation in its wider sense, and especially against those who equate conservation with the absence of human activity or interference. Much of this paper's content is based on my own experience as a water pollution control official and as Vice-President or Secretary of the Scottish Anglers Association for nearly 20 years.

2 The role of the angler

Angling has often been referred to as 'the gentle art' and in times past the image of the angler would have been of a man in accord with nature. In the many thousands of books which have been written by, and for, anglers, it is quite clear that the recreation is concerned with a great deal more than the capture of fish. The angler, if he was to be successful, had to understand a great deal about the aquatic environment and its inhabitants. The good angler would be able to identify at least some of the insect larvae and adults, he would know the seasons for emergence, and he would be aware of the inter-relationships between the various life forms of the stream or pond. In short, the angler would not have been thought of as merely a hunter of fish, but as a bit of a naturalist, interested in many of the aspects of the river or lake where he pursued his hobby, and a natural part of the country scene.

Times have changed, however, and in recent years angling and anglers have been in dispute with other water users. Anglers have protested about the activities of boaters, canoeists and water-skiers, and in turn have been subjected to criticism by those who see angling as detrimental to, or competing with, their own enjoyment of the waterways.

The anti-blood-sport league has branded angling as a cruel sport, a viewpoint that anglers are particularly sensitive to as it is at variance with their own image of themselves.

Research has shown that some of the practices of anglers are harmful not only to fish themselves, but to the riverine habitat in general, with vegetation and birdlife particularly damaged.

Anglers are now on the defensive, accused of direct cruelty to fish, indirect cruelty to other animals, vandalism to property, the thoughtless destruction of habitats, interference with breeding birds, and of littering the water and waterside with their rubbish and, in particular, the deadly by-products of their sport, lost and discarded tackle. As a result of press attention and reports in the general media, the public have become more aware of the issue recently, further increasing the 'anti-angling' feeling and leading to an isolation of anglers themselves.

The previous image of the angler as someone who moved in the countryside, in tune with nature, quietly pursuing his harmless hobby has been shattered. Was this previous image false? Has the angler changed in modern times (as his tackle has), and is there a lack of perspective among the critics so that faults are magnified and good points diminished or forgotten? In all these matters a sense of proportion has to be introduced and, in terms of acceptable social behaviour, the number of critics, as opposed to the numbers of those whose conduct is being criticized, must be weighed in the balance.

In a country where wildlife conservation is seen by many activists as an exercise in keeping other people out of their area of influence or interest, the present-day angler can easily be identified as a problem rather than as a benefit.

In the first place, there are so many of them. It is estimated that there may be over 3 million anglers in Britain, and as a participant sport it rates among the most popular. In terms of actual outings, it would appear that there are more angling engagements per year than attendances at professional football matches. This perhaps gives some idea of the demand for access to water that is being generated (Travis Commission 1981).

In addition, anglers are mobile – some are equipped and willing to travel long distances to practise their recreation – and they are willing to trek into remote areas on foot to fish a loch or stream picked from a map as a likely prospect. It is interesting to note here that in the Travis Commission report, anglers rated 'the chance to be alone in a quiet peaceful spot' as one of their main motivations for engaging in the sport.

Anglers have wide and varied tastes for different types of fish and different types of waterbody. Salmonids or coarse fish; individual specimens or catch measured in weight per hour; torrent, river, canal or estuary, lake or tarn, reservoir or gravel pit – anglers like them all. They can find interest and enjoyment in almost the full range of aquatic habitats available in this country.

Even if anglers were not highly mobile, they arise naturally from the local community in every part of the country. They are to be found in all socio-economic groups and in every trade and profession.

They are also represented in every age group – they tend to start angling before they are in their teens and are still fishing after they have retired from work, and from almost all other kinds of active sport.

Anglers are vocal, they are to a certain extent well organized, they form a significant political pressure group, and they can, if the need arises, generate large quantities of cash to purchase the right of access to fishing water.

It is perhaps hardly surprising, then, that various other groups with special interests in conservation have tended to distance themselves from the anglers, to criticize them, and to see them as competitors rather than allies. These differences are especially evident where the mode of operation of the conservation groups involves the reduction of human pressure. Anglers, in turn, are very sensitive to the possibility of tracts of water being denied to them, especially if historically they have had access. Several lochs and reservoirs in central Scotland have been closed to anglers either entirely or in part during the wildfowl breeding season.

The possibility of continuing and growing conflict to the detriment of both wildlife and humans is very real, and may only be reduced by a better understanding of the needs, contributions, and interdependence of all those concerned.

Having established that anglers and angling represent a major influence, for good or ill, on the freshwater habitat of this country, we may now consider what positive direction that influence has taken.

3 The contribution of anglers to wildlife conservation

If we accept that wildlife in fresh water includes fish, and that the whole pyramid, from the kingfisher (*Alcedo atthis*) downwards, depends on fresh water being able to support the small plants and animals which form the base of the productivity structure, then anglers have made a significant contribution to conservation in Britain.

River purification and, more recently, the control of pollution in still bodies of water have received tremendous encouragement from anglers over many years. Without their political involvement and pressure on a succession of Governments, the maintenance and, indeed, improvement of many of Britain's rivers and waterways would never have taken place. The angling political lobby did much to retain the independence of River Purification Boards in Scotland, and the water pollution control legislation throughout Britain would not have been drafted or enforced in its present form without that continuing involvement.

Quite apart from the development of legislation, River Purification Boards are very dependent on anglers to warn them of pollution incidents, some of which would otherwise go unnoticed (Clyde River Purification Board 1985). Sources of pollution of waterbodies in Britain can be subdivided in 3 ways. First, there are the discharges through pipes, licensed by the pollution control authorities, monitored by them, and probably controllable without outside help. Second, there are the incidents, amounting, in my own Board's area, to many hundreds each year, which involve an unexpected release of pollutants and which are normally detected by an observer on the river bank, often an angler. These incidents might include releases of silage liquor or slurry from farms, spillages of oil from ruptured storage tanks or vehicles, acts of vandalism involving the release of chemicals or oils from storage areas, illicit dumping, careless use of agricultural chemicals, including sheep-dip and other pesticides, and so on.

In Scotland at least, the vast majority of such incidents which are reported to the pollution control authorities are reported by anglers – very few by other members of the public. It is the knowledge that anglers are about and will report such incidents that stops many of our rivers being open drains. There is no doubt (in my mind) that, without the influence and concern of anglers, many of our rivers and estuaries would still be devoid of wildlife. As it is, the return of benthic fauna, then fish, then fish-eating birds and mammals to several of our most important rivers, including the Thames, the Clyde and the Tyne, owes much to the dedication and political pressure of anglers.

The third source of pollution presently threatening rivers and lakes is diffuse, such as the drainage of nutrients and pesticides from land, or acid rain from the atmosphere. These incidents have in some cases been highlighted in the first instance by anglers, who found that the fish or flies in which they were interested were disappearing, or were being affected in some other way. Loch Enoch and Loch Fleet in south-west Scotland provide good examples (Watt Committee on Energy 1984).

Many examples can also be given of anglers providing a very direct and local influence on conserving wildlife habitats (Tivy 1980; Gove 1985).

Anglers have had much involvement in ensuring that drainage channels, at their most efficient for water removal when absolutely straight and uniform in section, retained some of the characteristics of a natural river, with bends, bankside vegetation, pools and runs. However, it should also be stated that vegetation is often cut back by anglers to permit easier access for fishing (Swales & O'Hara 1983).

When hydro-electric schemes and dams for other purposes have been built, anglers have promoted moves to ensure the passage of fish, the release of compensation water to maintain a reasonable river flow downstream, and the avoidance of abrupt changes in water level in the reservoir, thus protecting a variety of wildlife forms (Pyefinch 1966).

The control of afforestation to protect rivers is also a major concern of anglers. They have done much to encourage a sensible drainage policy on hill slopes so that erosion does not fill the adjoining watercourses with silt. Forest owners have been persuaded to plant conifers back from the sides of streams and ponds to allow the natural fauna to thrive (Mills 1980).

Anglers have campaigned successfully to limit the effects of sand and gravel extraction, within the river channels, thereby protecting many species of sensitive insects, the spawning beds of salmonid fish, and the whole character of the river bed, which may be subject to substantial change when individual components are selectively removed. A good example is the stopping of sand and gravel extraction from the River Clyde in central Lanarkshire.

Recent legislation on small dams, involving the need for careful and expensive inspection by consulting engineers, has encouraged dam owners, including farmers and local authorities, to empty the dam and breach the containing walls, thus avoiding the expense of the surveys. However, the result has been a loss of desirable habitat for aquatic life, and anglers, through the Scottish Anglers Association, have been active in trying to ensure that such dams remain intact.

Many worked-out gravel pits and similar holes-in-the-ground, have been reclaimed as amenity lakes – principally for fish – but, as has been said previously, inevitably providing benefits and habitats for other forms of wildlife.

No doubt, there are other examples of how anglers are working towards wildlife conservation in the aquatic environment, and they should be in a position to be included within the conservation groups.

It would be a pity if the differences of viewpoint and competition for the privilege of managing some of this country's scarcest and most valuable resources result in conflict which would be harmful to the practice of conservation in the widest and most general sense of the word. It has to be accepted that all human conservation activities, no matter how altruistic they may appear, are for our pleasure and are essentially selfish in nature. Furthermore, there are very few of our conservation activities that do not interfere with the desires and aims of other people, so that conservation of wildlife is essentially restrictive and exclusive in human terms.

Education and a fuller appreciation of the consequences of careless action by anglers are required. The message should be delivered in as friendly a way as possible, and I feel sure that the principal angling organizations will do their utmost to promote good conduct and practice among their members.

Great Britain is a very small country, and there is a large and growing demand for access to, and use of, bodies of fresh water. Anglers view with concern the attitude of groups which seek by direct ownership, or by management agreements, to obtain effective control of freshwater habitats and then restrict human activity in favour of mammals, birds, and those few people deemed qualified to enjoy observing them. In some cases, this approach may be the only effective one for the long-term conservation of the wildlife, but there would appear to be scope for a balancing of interests so that the optimum benefit for people and wildlife can be derived from the management strategy.

Conservation has to be positive, but I would strongly recommend co-operation as opposed to confrontation, communication instead of recrimination, and an appreciation of other points of view.

4 Summary

Angling and wildlife conservation are increasingly being identified as in conflict with each other. The angler, as seen through the eyes of some other enjoyers of the waterside environment, can be intrusive, damaging, insensitive and, at times, vandalistic. Anglers are becoming a focus of attention for the anti-blood-sport movement, and there is increasing antagonism between anglers and those who seek to promote other forms of life, in particular birds, associated with water. The main problem with anglers is that there are so many of them. As individuals and groups, they have done much to conserve our waterways and associated fauna and flora. Much of the general criticism levelled against them is unfair to the vast majority, and this paper sets out to say something of the favourable impacts which anglers have on 'conservation'.

5 References

Clyde River Purification Board. 1985. *Annual report.* East Kilbride: Clyde River Purification Board.

Gove, A. J. 1985. *The restoration of rivers and streams.* Boston: Butterworth.

Mills, D. H. 1980. *The management of forest streams.* Mansfield: HMSO.

Pyefinch, K. A. 1966. Hydro-electric schemes in Scotland, biological problems and effects on salmonid fisheries. In: *Man-made lakes,* edited by R. H. Lowe-McConnell, 139–145. London: Institute of Biology & Academic Press.

Swales, S. & O'Hara, K. 1983. A short-term study of the effects of a habitat improvement programme on the distribution and abundance of fish stocks in a small lowland river in Shropshire. *Fish. Manage.,* 14, 135–144.

Tivy, J. 1980. *The effect of recreation on freshwater lochs and reservoirs in Scotland.* Perth: Countryside Commission for Scotland.

Travis Commission. 1981. *Angling in Britain 1980.* London: Angling Foundation.

Watt Committee on Energy. 1984. *Acid rain.* (Report no. 14.) London: Watt Committee on Energy.

Angling and wildlife conservation – are they incompatible?

P S MAITLAND* and A K TURNER†
*Institute of Terrestrial Ecology, Bush Estate, Penicuik
†Department of Zoology, University of Glasgow, Glasgow

1 Introduction

The publication of which this paper is a part is based on a series of papers presented at a joint meeting organized by the Scottish Freshwater Group and the British Ecological Society in October 1985. Those invited to contribute to the meeting were asked to confine themselves to the theme of angling and wildlife in fresh water; otherwise, no restrictions were placed on the presentations. As a result, the approaches to the topic have varied: some are based on recent unpublished work, others are reviews of the state of knowledge, yet others represent statements of opinion rather than presentation of facts. These attitudes reflect the, sometimes controversial, public debate in which conservationists produce most of the data and anglers express their strongly held, but often less factual, opinions. All the papers have been edited and commented upon by independent referees (to whom we are very grateful), but the final texts are those of the authors and do not necessarily reflect the views of the editors, the Scottish Freshwater Group or the British Ecological Society.

The evidence presented would seem to indicate that the activities of anglers do harm certain kinds of wildlife in some situations. The conflicts have been reviewed by Edwards and Bell (1986). Much of the current controversy has arisen over the impact of lead fishing weights on swans (*Cygnus olor*) in England (Thomas *et al.* 1987), and there is no doubt that this is a serious problem in some areas. Anglers' litter, in addition to being unsightly and present in surprisingly large amounts at popular fisheries (Edwards & Cryer 1987), also has a serious impact on some birds and mammals. The presence of anglers often disturbs birds (Cooke 1987) and mammals, but, in some cases at least (eg otters (*Lutra lutra*)), this disturbance may not always be as serious as often supposed (Jefferies 1987). The activities of anglers can also alter the habitat in various ways (Murphy & Pearce 1987), sometimes unintentionally (eg by trampling down vegetation), but often intentionally as a management procedure (eg weed cutting and bank clearance). Anglers may also impinge directly on the biological communities involved by poisoning unwanted fish species (Morrison 1987), shooting supposed predatory birds (Mills 1987) or carelessly introducing new species to a water by releasing live bait at the end of a day (Maitland 1987).

Anglers respond to the accusations of harming wildlife in a number of ways. On the positive side, they point out that on a number of waters satisfactory multipurpose use has been achieved where anglers, birdwatchers and other groups of people appear satisfied with the situation (Parry 1987). In addition, partly because they are so numerous and therefore a powerful lobby, anglers also help to conserve natural waters by opposing pollution and other aspects of environmental damage (Mackay 1987). They are also an important element in detecting serious pollution incidents at an early stage and warning the authorities accordingly. On the other hand, some anglers deny that there is any evidence of damage. Others place the blame on a small unrepresentative minority of the angling fraternity, and point out also that there are many other types of water users creating similar problems – litter, disturbance, pollution, habitat damage, community modification, and so on.

It is the purpose of this paper to review the real problems in this conflict and to suggest ways in which they could be solved to the mutual satisfaction of both anglers and conservationists.

2 Problems and solutions

The increasing evidence of damage to swans and other wildfowl from lead fishing weights (and attached nylon) lost or disposed of by anglers makes it quite unacceptable that the situation can continue, and indeed considerable effort has gone into finding acceptable alternatives to lead. These alternatives are now available and there seems no reason why the manufacture, sale and use of the original dangerous weights should not stop, preferably voluntarily, but more realistically through appropriate legislation (which is at present receiving approval). How rapidly the environment will recover from such a ban is uncertain for, whilst past accumulations of lead shot are likely to remain for decades, centuries or even millennia, their accessibility for wildlife may be more limited in time.

Litter is a common problem in human society, always involving unsightliness and sometimes danger to wildlife and humans. Much angling litter is commonplace (eg waste paper, food cartons) but some is quite characteristic of the culprit (eg bait cans, hooks and nylon line). Angling is, to some extent, an organized sport, and there seems every reason to suppose that, if the will is there on the part of both angler and site manager, litter could be controlled. In some instances, more thought needs to be given to site design and maintenance, and further efforts devoted to the development of angler tackle, such as line which, once discarded, is recovered easily or degraded naturally.

Although anglers do disturb birds and mammals (and fish!), so do many other waterside users, sometimes to an even greater extent where large numbers of people are involved. However, that should not be an excuse for anglers to evade their responsibilities. The onus is partly on them and partly on other interested groups to safeguard important wildlife areas or species by restricting their activities to the least vulnerable areas. Whilst there may be aspects of behaviour which distinguish the angler from other groups, perhaps the onus in safeguarding important wildlife areas or species lies with the conservationists, in restricting access and directing the public in general, as well as anglers in particular, to the least vulnerable areas.

Much habitat modification, however, lies in the hands of the managers of fisheries. Where the damage is indiscriminate (eg damage to vegetation by trampling or boats), there may need to be some restrictions. In addition, in considering bank clearance or river engineering of any kind, much more thought could be given to the needs of wildlife (eg the importance of cover to otters on at least one bank). The suitability of an area for different types of fishing also needs to be considered. Boat fishing, for example, may be more acceptable than bank fishing at a particular site because it would cause less disturbance to breeding and feeding birds.

The control of unwanted fish species by poisoning, and bird and mammal predators by shooting, once an entirely acceptable part of fishery management, has caused considerable controversy in recent years. The ecological basis for such practices is now being questioned, and a considerable amount of research needs to be carried out in the field of population dynamics, and particularly in such areas of competition as predator-prey relationships, before the efficiency of these practices can be rigorously assessed. The wholesale destruction of entire fish populations by piscicides simply to replace them with favoured (usually salmonid) species should also be considered more carefully than at present, taking into account all the relevant wildlife factors, including the needs of rare fish species. As well as removing fish from systems, anglers have been one of the main factors in the dispersal of fish species within the British Isles and here, too, much more control is needed if important systems and communities are to be adequately protected.

The idea of fish being regarded not only as the angler's quarry but also as wildlife requiring protection (Maitland 1985) may be taken further, and damage to fish stocks may be caused even when fish are returned to the water after capture. Some angling practices (eg keep-nets and weighing-in procedures) may adversely affect individual populations which are heavily fished. Groundbaiting may also have profound local effects on benthic communities, even though its contribution to the budgets of organic matter or phosphorus at sites is rarely substantial (Edwards & Fouracre 1983).

3 A code of conduct

Further legislation is often put forward as the best way of dealing with many of the kinds of problems discussed above. In fact, many of them are already covered by existing legislation (eg litter disposal, use of piscicides, introductions of fish) and, although this legislation could (and should) be improved and strengthened in some areas, another useful (and parallel) approach is through the development of a wider awareness among those involved in the problem. In the case of angling, the development of a code of conduct should be undertaken by the main national angling bodies. Unfortunately, at present, only a minority of anglers are associated with a club or organization, a situation that needs to be changed if any code of conduct is to be widely adopted. Perhaps more anglers would be willing to join organizations if more of the 'best' angling sites were controlled by (and therefore the responsibility of) local angling associations. Some of the most important suggested items considered in such a code could be as follows.

i. Materials used by anglers should be as harmless to wildlife as possible. In particular, the use of lead should be completely phased out and, where possible, degradable lines, etc, should be developed.
ii. Anglers should be strongly encouraged to take all litter home with them or dispose of it safely. Those responsible for managing angling sites and organizing competitions should provide litter receptacles, where appropriate, and penalize anglers responsible for littering.
iii. Anglers should be encouraged to be more aware of the damage certain practices cause to habitats and to avoid such activities. Those responsible for managing sites should consult conservationists about the wildlife value of angling sites, and in particular about the presence of sensitive species. They should also take appropriate steps to minimize damage, for example by restricting access to sensitive areas and protecting eroding banks. Those involved in direct habitat manipulation for angling should seek advice from conservation bodies at the planning stage. More involvement of anglers with conservation issues would also be helpful, eg by having an angling representative associated with the local Naturalists' Trust and a conservation representative on the angling club committee.
iv. Open discussion of the facts (or lack of them) concerning the role and impact of fish, avian and mammalian predators should be encouraged among anglers. Control measures involving the use of piscicides or the shooting of birds should be used only after careful consideration. Alternative measures should be considered more widely.
v. All anglers should be made aware of the law relating to the movement of fish into and within Great Britain. Indiscriminate introductions of fish should be discouraged. Species of fish should not be used as live bait.

Plate 15. Loch Lomond – an internationally famous and successful multi-purpose water, whose fishery is well controlled by an angling association with over 1000 members. Part of the loch is a National Nature Reserve, and it is also important for water supply, boating, water-skiing, bathing and as an amenity (Photograph P S Maitland)

vi. There are more than 50 000 lakes in Great Britain (Smith & Lyle 1979) and a similar number of streams and rivers, yet much of this extensive resource is under pressure. Anglers should be encouraged to remember that many other types of water users have requirements, and that priorities vary from place to place. Some waters are best suited for angling only, at others a multi-purpose use (eg angling, water supply, bird-watching) may be feasible (Plate 15), at yet other, perhaps important conservation sites, there should be no angling. There should be more consideration of zoning usage of sites, by space and/or time, to minimize or avoid damage to wildlife. Angling sites should be designed to encourage and channel the activities of anglers. Some important conservation waters have, and should continue to have, no fish, and their intrinsic scientific value should be made clear.

vii. A continuing problem in developing management plans for fresh waters which involve anglers is the lack of adequate information on fish catches and effort. Where good data are available, they provide a rational basis for the management of the fishery (Bailey-Watts & Maitland 1984) and its integration with other local interests (Plate 16). Angling organizations and those responsible for the management of fisheries should be encouraged to develop and adopt better methods of collecting and storing information on anglers and their catches in all parts of the country.

viii. Angling organizations should seek further to cultivate a responsible attitude to the environment among all anglers, attract more anglers into joining an organization, and be more prepared to censor anglers who are irresponsible.

4 Research requirements

It is evident from the papers presented at the meeting, and from the ensuing discussion (which was not recorded), that not all the possible effects of angling on wildlife in fresh waters are fully known. Although many questions remain to be answered, that is no excuse for inaction on the part of the angling community. The effects of lead on birds are incontrovertible and voluntary action or legislation is needed now. The effects of litter, disturbance, habitat modification, piscicides and fish introductions are clear enough, even if more research is desirable. As yet, the merits of predator control are unclear, and this is a most important area for investigation. Thus, scientific research into several aspects of the problem is needed.

i. There is an important gap in our knowledge of predator-prey relationships. What effect do pre-

Plate 16. Loch Leven – a National Nature Reserve of international importance for its wildfowl, but also a world-renowned and successful trout fishery. Good control of the fishery has meant that catch data are available for more than a century and these data have allowed a rational basis for fishery management (Photograph P S Maitland)

dators have on fish populations? Do they take surplus young fish without reducing the numbers available for angling? Do they, in any case, take unhealthy, perhaps diseased or parasitized, fish, leaving healthier ones for the anglers? Lastly, if predators are causing serious losses, how can they be effectively discouraged from taking fish, instead of being shot? Killing predators is often costly and at best only a short-term measure. Habitat manipulation to make fish less available may be a more cost-effective long-term solution.

ii. More studies of the effects of angling on potentially sensitive and rare species are needed, particularly of the long-term effects on breeding and feeding behaviour, rather than just on the animal's initial reaction to humans. The behaviour of common species, such as mallard (*Anas platyrhynchos*), may not be indicative of the reactions of rarer wildfowl, for example. Most emphasis has been placed on birds so far, but animals other than birds need to be studied too. The level of disturbance tolerable at a site may depend on the species it is wished to encourage there.

iii. Much is known about the effects of habitat modification on wildlife, but there are gaps in our knowledge. For example, how does aeration affect the growth of aquatic plants? How can emergent plants be encouraged to grow along banks disturbed by boats?

iv. Safer alternatives to the present piscicides are needed to remove unwanted fish. We also need to know much more about the effects of competition among fish species. Is it really necessary to remove the unwanted fish already present at a site before introducing other, more favoured, species? What is the effect of introducing a new species to an existing stable community?

v. More also needs to be known about the effects of angling practices on fish populations themselves, for they too are wildlife. Thus, we need to know more about the impact of capture, keepnets, weighing-in procedures at competitions, and groundbaiting on fish populations and their environment.

vi. It is also important to know more about the behaviour and attitudes of anglers so that acceptable and cost-effective solutions to various problems can be found.

5 Conclusions

Despite the obvious conflicts of interest in some areas, angling and wildlife are not necessarily incompatible. Where due consideration can be given to the effects of litter, habitat modification, disturbance and control of predators and competitors of fish, the 2 can often co-exist harmoniously. It may be, however, that at some sites priority is given to the anglers and at others to the wildlife, so that over a large area both can survive, if not actually co-exist.

Before such a situation is possible, however, more consideration of wildlife interests by anglers is needed. In particular, the sport needs to be better organized so that more anglers, by being members of angling clubs, are controlled to some extent, can be made aware of the needs of wildlife more effectively, and can be penalized if they are irresponsible. In England and Wales, control should be easier because all anglers need licences to fish. At present, communication amongst anglers is difficult because many do not belong to clubs or other angling bodies. In this connection, the numerous angling journals that are available could help enormously by carrying articles on various conservation issues from time to time. Angling organizations and site managers need to be more aware of wildlife, especially when designing angling sites and planning the modification of habitats or the control of mammals, birds or fish so that the effects can be minimized. In turn, conservationists can work more closely with anglers so that the latter are not kept out of areas where they would have a negligible effect on the wildlife.

6 Summary

This paper reviews the main conflicts between angling and wildlife conservation in fresh waters. The major problems which have arisen relate to the use of lead fishing weights, litter disposal, disturbance, habitat alteration, the use of piscicides, shooting of bird and mammal predators, and the introduction and translocation of fish species. On the other hand, anglers are said to be very beneficial in supporting controlled multi-purpose use of waters and acting as a powerful lobby for pollution prevention. A code of conduct for anglers is suggested which would, to a large extent, eliminate the harmful effects of their activities. In addition, a series of research topics is proposed which would give answers to some of the problems that exist in this field at present.

7 Acknowledgements

We are most grateful to Professor R W Edwards, Mr P H Tombleson and Dr M Marquiss for constructive criticism of early drafts of this paper.

8 References

Bailey-Watts, A. E. & Maitland, P. S. 1984. Eutrophication and fisheries in Loch Leven, Kinross, Scotland. *Proc. Inst. Fish. Mgmt Ann. Study Course*, **15**, 170–190.

Cooke, A. S. 1987. Disturbance by anglers of birds at Grafham Water. In: *Angling and wildlife in fresh waters*, edited by P. S. Maitland & A. K. Turner, 15–22. (ITE Symposium no. 19.) Grange-over-Sands: Institute of Terrestrial Ecology.

Edwards, R. W. & Bell, D. 1986. The impact of angling on wildlife. *Proc. Br. Freshwat. Fish. Conf.*, **4**. In press.

Edwards, R. W. & Cryer, M. 1987. Angler litter. In: *Angling and wildlife in fresh waters*, edited by P. S. Maitland & A. K. Turner, 7–14. (ITE Symposium no. 19.) Grange-over-Sands: Institute of Terrestrial Ecology.

Edwards, R. W. & Fouracre,V. A. 1983. Is the banning of groundbaiting in reservoirs justified? *Proc. Br. Freshwat. Fish. Conf.*, **3**, 89–94.

Jefferies, D. J. 1987. The effects of angling interests on otters, with particular reference to disturbance. In: *Angling and wildlife in fresh waters*, edited by P. S. Maitland & A. K. Turner, 23–30. (ITE Symposium no. 19.) Grange-over-Sands: Institute of Terrestrial Ecology.

Mackay, D. W. 1987. Angling and wildlife conservation. In: *Angling and wildlife in fresh waters*, edited by P. S. Maitland & A. K. Turner, 72–75. (ITE Symposium no. 19.) Grange-over-Sands: Institute of Terrestrial Ecology.

Maitland, P. S. 1985. Criteria for the selection of important sites for freshwater fish in the British Isles. *Biol. Conserv.*, **31**, 335–353.

Maitland, P. S. 1987. Fish introductions and translocations – their impact in the British Isles. In: *Angling and wildlife in fresh waters*, edited by P. S. Maitland & A. K. Turner, 57–65. (ITE Symposium no. 19.) Grange-over-Sands: Institute of Terrestrial Ecology.

Mills, D. H. 1987. Predator control. In: *Angling and wildlife in fresh waters*, edited by P. S. Maitland & A. K. Turner, 53–56. (ITE Symposium no. 19.) Grange-over-Sands: Institute of Terrestrial Ecology.

Morrison, B. R. S. 1987. Use and effects of piscicides. In: *Angling and wildlife in fresh waters*, edited by P. S. Maitland & A. K. Turner, 47–52. (ITE Symposium no. 19.) Grange-over-Sands: Institute of Terrestrial Ecology.

Murphy, K. J. & Pearce, H. G. 1987. Habitat modification associated with freshwater angling. In: *Angling and wildlife in fresh waters*, edited by P. S. Maitland & A. K. Turner, 31–46. (ITE Symposium no. 19.) Grange-over-Sands: Institute of Terrestrial Ecology.

Parry, M. L. 1987. Multi-purpose use of waters. In: *Angling and wildlife in fresh waters*, edited by P. S. Maitland & A. K. Turner, 66–71. (ITE Symposium no. 19.) Grange-over-Sands: Institute of Terrestrial Ecology.

Smith, I. R. & Lyle, A. A. 1979. *Distribution of freshwaters in Great Britain*. Cambridge: Institute of Terrestrial Ecology.

Thomas, G. J., Perrins, C. M. & Sears, J. 1987. Lead poisoning and waterfowl. In: *Angling and wildlife in fresh waters*, edited by P. S. Maitland & A. K. Turner, 5–6. (ITE Symposium no. 19.) Grange-over-Sands: Institute of Terrestrial Ecology.

Appendix: List of participants

Mrs S M Adair, Institute of Terrestrial Ecology, Bush Estate, Penicuik, EH26 0QB.

Mr C Adams, University Field Station, Rowardennan, Glasgow, G63 0AW.

Miss J Allan, University Field Station, Rowardennan, Glasgow, G63 0AW.

Mr J Armstrong, Department of Zoology, University of Aberdeen, Tillydrone Avenue, Aberdeen, AB9 2TN.

Miss K Atkinson, Freshwater Biological Association, Ambleside, LA22 0LP.

Mr N K Atkinson, Tadorna, 5 Tolmount Crescent, Montrose.

Mr C Badenoch, Nature Conservancy Council, Galahill House, Barr Road, Galashiels.

Ms S Bailey, 15 MacDowall Road, Edinburgh, EH9 3ED.

Dr A E Bailey-Watts, Institute of Terrestrial Ecology, Bush Estate, Penicuik, EH26 0QB.

Mr P J Barrett, 1 St Andrews Drive, Bridge of Weir.

Mr D M Bayne, Nature Conservancy Council, Airthrey Castle, Stirling, FK9 4LA.

Dr J Berry, Tayfield, Newport-on-Tay, DD6 8HA.

Mr C M Booth, 27 Viewforth, Edinburgh.

Mr P Bowden-Smith, Loch Awe Improvement Association, Hayfield, Kilchrenan, Taynuilt.

Mr D Brown, University Field Station, Rowardennan, Glasgow, G63 0AW.

Mr D M Bryden, Drumpellier Country Park, Townhead Road, Coatbridge.

Ms A Burch, Top Flat, 98 Marchmont Crescent, Edinburgh.

Dr R N B Campbell, Institute of Terrestrial Ecology, Bush Estate, Penicuik, EH26 0QB.

Mr D Carss, Forest Cottage, Ardmaddy, Oban, PA34 4QY.

Mrs H Cattan, University Field Station, Rowardennan, Glasgow, G63 0AW.

Mr B Cleator, 87 Spottiswoode Street, Edinburgh, EH9 1BZ.

Mr B Clelland, Forth Road Purification Board, West Mill Road, Colinton, Edinburgh, EH13 0NX.

Mr A W Colling, Biel Mill, Dunbar.

Dr A S Cooke, Nature Conservancy Council, Northminster House, Peterborough, PE1 1UA.

Ms C Crawford, Nature Conservancy Council, Airthrey Castle, Stirling, FK9 4LA.

Dr M Cryer, Department of Applied Biology, UWIST, Cardiff, CF1 3XF.

Mr M B Davidson, North East River Purification Board, Persley, Aberdeen, AB2 2UQ.

Mr D Dick, Royal Society for the Protection of Birds, 17 Regent Terrace, Edinburgh, EH7 5BN.

Ms J Driver, 2 Barony Street, Edinburgh, EH3 6PE.

Dr A Duncan, Department of Zoology, University of London, Englefield Green, TW20 9TY.

Mr D A Dunkley, Freshwater Fisheries Laboratory, 6 California Street, Montrose, DD10 8DW.

Mr K East, Institute of Terrestrial Ecology, Bush Estate, Penicuik, EH26 0QB.

Professor R W Edwards, Department of Applied Biology, UWIST, Cardiff, CF1 3XF.

Mr B P Emanuel, University Field Station, Rowardennan, Glasgow, G63 0AW.

Mr I Fozzard, Forth River Purification Board, Alloa Road, Stirling, FK9 5PN.

Mr J Frame, Clyde River Purification Board, Murray Road, East Kilbride, G75 0LA.

Mr A Fuller, Scottish Marine Biological Association, Dunstaffnage, Oban, PA34 4AD.

Mr C A Galbraith, University of Aberdeen, Culterty Field Station, Newburgh, AB4 0AA.

Miss N J Gordon, Nature Conservancy Council, 12 Hope Terrace, Edinburgh, EH9 2AS.

Mr P Geddes, Freshwater Fisheries Laboratory, Pitlochry, PH16 5LB.

Mr E Gillespie, Freshwater Fisheries Laboratory, Pitlochry, PH16 5LB.

Dr J J D Greenwood, Department of Biological Sciences, University of Dundee, Dundee, DD1 4HN.

Mr R B Greer, Freshwater Fisheries Laboratory, Pitlochry, PH16 5LB.

Mr A K Haddison, 11 Greenbank Terrace, Edinburgh.

Mr J S Halloran, Department of Zoology, University College, Cork.

Mr J D Hamilton, Department of Biology, Paisley College of Technology, High Street, Paisley, PA1 2BE.

Mr G Hanslip, Freshwater Fisheries Laboratory, Pitlochry, PH16 5LB.

Ms M Harding, 15 MacDowall Road, Edinburgh, EH9 3ED.

Miss P Harvey, University Field Station, Rowardennan, Glasgow, G63 0AW.

Mr D W Hay, Freshwater Fisheries Laboratory, Pitlochry, PH16 5LB.

Dr A D Hawkins, Marine Laboratory, Victoria Road, Aberdeen.

Dr C Henty, 7 Coney Hill, Bridge of Allan.

Mr P Higson, Drumpellier Country Park, Townhead Road, Coatbridge.

Mr A V Holden, Achnasithe, Manse Road, Moulin, Pitlochry, PH16 5EP.

Mr R G Hollister, Auchindoun, Bank End Road, Bridge of Weir, PA11 3EU.

Mr E T Hunter-Blair, Parton House, Castle Douglas, DG7 3NB.

Mr A K M S Islam, University Field Station, Rowardennan, Glasgow, G63 0AW.

Mr A B Jamieson, Lothian Regional Council, 12 St Giles Street, Edinburgh, EH1 1TU.

Dr D J Jefferies, Nature Conservancy Council, Northminster House, Peterborough, PE1 1UA.

Mr R Johnston, 29 Lauder Road, Edinburgh.

Mr D H Jones, Institute of Terrestrial Ecology, Bush Estate, Penicuik, EH26 0QB.

Dr P J Jones, Department of Forestry and Natural Resources, University of Edinburgh, Mayfield Road, Edinburgh, EH9 3JU.

Mr S Justice, Department of Biological Sciences, Napier College, Colinton Road, Edinburgh, EH9 2DT.

Miss S Lagden, University Field Station, Rowardennan, Glasgow, G63 0AW.

Mr D J Lampard, Department of Biological Sciences, Glasgow College of Technology, Cowcaddens Road, Glasgow, G4 0BA.

Mr E D LeCren, High Point, Roman Road, Appleby, CA16 6JB.

Ms M C Lindsay, British Waterways Board, Applecross Street, Glasgow, G4 9SP.

Dr C Lindsey, Department of Applied Science, Brighton College of Technology, Pelham Street, Brighton, BN1 4FA.

Mr D Lowson, Forth River Purification Board, West Mill Road, Colinton, Edinburgh, EH13 0NX.

Mr A A Lyle, Institute of Terrestrial Ecology, Bush Estate, Penicuik, EH26 0QB.

Mr A MacFarlane, University Field Station, Rowardennan, Glasgow, G63 0AW.

Mr R McFarlane, 18A Carlton Terrace, Edinburgh, EH7 5DD.

Mr D W Mackay, Clyde River Purification Board, Murray Road, East Kilbride, G75 0LA.

Mr J Mackenzie, Atlantic Salmon Trust, Druimuan, Killiecrankie, Pitlochry, PH16 5LG.

Mr W D McKenzie, Highlands and Islands Development Board, Bank Street, Inverness, IV1 1QR.

Mr Q McLaren, Borders Regional Council, Newton St Boswells, TD6 0SA.

Mr M J McLaughlin, 21 Kinross Avenue, Glasgow, G52 3JB.

Dr D S McLusky, Department of Biological Science, University of Stirling, Stirling, FK9 4LA.

Mr R McMath, University Field Station, Rowardennan, Glasgow, G63 0AW.

Miss S McMath, University Field Station, Rowardennan, Glasgow, G63 0AW.

Dr P S Maitland, Institute of Terrestrial Ecology, Bush Estate, Penicuik, EH26 0QB.

Dr M Marquiss, Institute of Terrestrial Ecology, Bush Estate, Penicuik, EH26 0QB.

Mr M W Marsden, Forth River Purification Board, Alloa Road, Stirling, FK9 5PN.

Dr L May, Institute of Terrestrial Ecology, Bush Estate, Penicuik, EH26 0QB.

Mr I A D Millar, Remony, Aberfeldy, PH15 2HR.

Ms K Miller, North East River Purification Board, Persley, Aberdeen, AB2 2UQ.

Dr D H Mills, Department of Forestry and Natural Resources, University of Edinburgh, Mayfield Road, Edinburgh, EH9 3JU.

Dr H Milne, University of Aberdeen, Culterty Field Station, Newburgh, AB4 0AA.

Dr C I Morgan, Department of Biological Sciences, Glasgow College of Technology, Cowcaddens Road, Glasgow, G4 0BA.

Mr K H Morris, Institute of Terrestrial Ecology, Bush Estate, Penicuik, EH2 0QB.

Mr A W Morrison, Forth River Purification Board, 6 Acorn Court, Glenrothes, KY7 5LZ.

Mr B R S Morrison, Freshwater Fisheries Laboratory, Pitlochry, PH16 5LB.

Mr A Mowat, 41 Mayfield Road, Edinburgh, EH9 2NQ.

Dr K J Murphy, Department of Botany, University of Glasgow, Glasgow, G12 8QQ.

Mr R H D Orr, 31 Clerkington Road, Haddington.

Dr R Owen, North East River Purification Board, Persley, Aberdeen, AB2 2UQ.

Dr C J Palmer, Herbert Art Gallery and Museum, Jordan Well, Coventry, CV1 5RW.

Mr M L Parry, Severn Trent Water Authority, 2297 Coventry Road, Birmingham.

Dr H G Pearce, Lee Valley Regional Park Authority, Myddleton House, Bulls Cross, Enfield, EN2 9HG

Dr C M Perrins, Department of Zoology, University of Oxford, Parks Road, Oxford, OX1 3PS.

Dr M J Picken, Scottish Marine Biological Association, Dunstaffnage, Oban, PA34 4AD.

Dr K B Pugh, North East River Purification Board, Persley, Aberdeen, AB2 2UQ.

Mr R Raynor, Top Flat, 36 Hillside Crescent, Edinburgh, EH7 5EF.

Dr P Read, Department of Biological Sciences, Napier College, Colinton Road, Edinburgh, EH9 2DT.

Miss S Roma, University Field Station, Rowardennan, Glasgow, G63 0AW.

Mr A J Rosie, Highland River Purification Board, Dingwall, IV5 9QY.

Mr J Sage, Dundee City Museums and Art Galleries, Albert Square, Dundee, DD1 1DA.

Mr G S Scott, 61 King Street, Stonehouse, ML9 3EQ.

Mr S D Sedgwick, Fisherman's Hill, Comrie, PH6 2LX.

Mr C Self, 98 Marchmont Crescent, Edinburgh.

Mr W M Shearer, Freshwater Fisheries Laboratory, 6 California Street, Montrose, DD10 8DW.

Dr C B Shedden, Marford Mill, Rossett, Wrexham, LL12 0HL.

Dr J C Sheldon, Lothian Regional Council, 12 St Giles Street, Edinburgh, EH1 1TU.

Dr R G J Shelton, Freshwater Fisheries Laboratory, Pitlochry, PH16 5LB.

Mr T Sherratt, Top Flat Right, 61 Arden Street, Edinburgh.

Mr I R Smith, Institute of Terrestrial Ecology, Bush Estate, Penicuik, EH26 0QB.

Ms S Smith, 18 Dundonald Street, Edinburgh, EH3 6RY.

Ms C Sommerville, Institute of Aquaculture, University of Stirling, Stirling, FK9 4LA.

Dr A B Stephen, Institute of Aquaculture, University of Stirling, Stirling, FK9 4LA.

Mr C Steven, Department of Zoology, University of Durham, South Road, Durham, DH1 3LE.

Mr I Stirling, 18A Carlton Terrace, Edinburgh, EH7 5DD.

Dr I M Strachan, Lumbo Farmhouse, St Andrews, KY16 8NS.

Mr G Struthers, Freshwater Fisheries Laboratory, Pitlochry, PH16 5LB.

Mr O Tannahill, Easter Sheardale, Dollar.

Dr I R Taylor, Department of Forestry and Natural Resources, University of Edinburgh, Mayfield Road, Edinburgh, EH9 3JU.

Dr K Taylor, British Trust for Ornithology, Beech Grove, Tring, HP23 5NR.

Ms A Thomson, 120 Buccleuch Street, Edinburgh, EH8 9NQ.

Mr M W Thomson, Anglers' Co-operative Association, 21 Heriot Row, Edinburgh, EH3 6EN.

Dr W Thorburn, 5 The Cottages, Craigton, Milngavie, G62 7HQ.

Dr J E Thorpe, Freshwater Fisheries Laboratory, Pitlochry, PH16 5LB.

Mr G E D Tiley, Agronomy Department, West of Scotland Agricultural College, Auchencruive, KA6 5HW.

Dr R Tippett, Department of Zoology, University of Glasgow, Glasgow, G12 8QQ.

Dr A K Turner, Department of Zoology, University of Glasgow, Glasgow, G12 8QQ.

Mr A F Walker, Freshwater Fisheries Laboratory, Pitlochry, PH16 5LB.

Mr N M Walker, The Vicarage, Hellifield, Skipton.

Mr D Weir, 26 Burlington Avenue, Glasgow.

Mr I A Wilson, Flat 4, 45 St Stephens Street, Edinburgh, EH3 5AH.

Mrs M S Wilson, 3 Howden Hall Gardens, Edinburgh, EH16 6UN.

Dr B J B Wood, Applied Microbiology Division, University of Strathclyde, George Street, Glasgow, G1 1XW.

Mr P Wright, University Field Station, Rowardennan, Glasgow, G63 0AW.

Mrs S Wurthmann, Department of Biological Sciences, Glasgow College of Technology, Cowcaddens Road, Glasgow, G4 0BA.

Mr L Yaoming, 12 Perth Street, Edinburgh, EH3 5DP.

Dr R E Youngman, 20 East Moulin Road, Pitlochry, PH16 5HY.

Mr X Zhu, Department of Forestry and Natural Resources, University of Edinburgh, Mayfield Road, Edinburgh, EH9 3JU.